THINKING ABOUT DOOM

THINKING ABOUT DOOM

A PHILOSOPHICAL INVESTIGATION INTO THE DOOMSDAY ARGUMENT

SASHA ZOUEV

BIRKBECK COLLEGE, UNIVERSITY OF LONDON
MRES PHILOSOPHY 2022
DISSERTATION

ZOUEV PUBLISHING
NW8 9PA, London
www.zouevpublishing.com

ISBN 978-1-9163451-6-4

First published as part of a Master of Philosophy postgraduate dissertation thesis at the New College of the Humanities, London

Copyright © Sasha Zouev 2023
This book is printed on acid-free paper.

Contact:
Alexander (Sasha) Zouev
Alexander.Zouev@gmail.com
0044 7710836424

TABLE OF CONTENTS

ABSTRACT: 8

I. INTRODUCTION 10

1.0 INTRODUCTION TO THE DA 10
1.1 GOTT'S DA 13
1.2 LESLIE'S DA 24

II. DOOMSDAY ARGUMENT THOUGHT-EXPERIMENTS 35

2.0 INTRODUCTION TO DA THOUGHT-EXPERIMENTS 35
2.1 LESLIE'S URNS 37
2.2 CONCEPTUAL ISSUE IN LESLIE'S URNS: REFERENCE CLASS 38
2.3 ECKHARTD'S SHOOTING ROOM 48
2.4 CONCEPTUAL ISSUE IN ECKHARDT'S SHOOTING ROOM: DETERMINISM 51
2.5 NORTHCOTT'S ASTEROIDS 57

2.6 CONCEPTUAL ISSUE IN NORTHCOTT'S ASTEROIDS: TRUMPING 60

III. EMPIRICAL TESTING OF THE DA 67

3.0 SOBER'S EMPIRICAL CRITICISM OF THE DA 67

3.1 SOBER'S OBJECTION TO GOTT 67

3.2 SOBER'S OBJECTION TO LESLIE 79

3.3 TESTING THE DA ITSELF VIA SIMULATION 91

IV. IMPLICATIONS OF THE DOOMSDAY ARGUMENT 103

V. CONCLUSION 106

REFERENCES 109

EXAMINER FEEDBACK 125

ABSTRACT:

The Doomsday Argument (DA) aims to give a prediction on humankind's longevity by taking into account how many humans have already existed. To this day, the prevailing consensus amongst those actively writing about the DA is that it still remains unrefuted, despite innumerable attempts in the literature to debunk it. This dissertation provides a critical assessment of the use of thought-experiments in DA rebuttals and argues that testing the DA empirically could yield more fruitful results.

We have split our paper into four parts. Part I introduces the reader to the DA, with an exposition of both Leslie's and Gott's DA. We accompany our explanations with detailed mathematical examples. This is followed by a discussion of why the argument is worth our philosophical interest.

In Part II, a critical assessment of the use of thought-experiments in DA literature is given with a focus on three distinct examples: Leslie's (1989a) Urns, Eckhardt's (1997) Shooting Room and Northcott's (2016) Asteroids. We argue that while these thought-experiments are useful in illuminating different issues associated with the DA (in our three respective examples those are reference class, determinism, and trumping) they ultimately cannot get us any closer to rejecting or accepting the DA. To have a realistic chance of doing so, we need actual testing of empirical data.

Part III provides a critical assessment of Sober's (2003) claim that testing the sampling assumptions underlying the DA can empirically disconfirm the DA. We argue that Sober's critique is valid when considering Gott's DA, however it is inconclusive with regards to Leslie's DA. Further, we contend that this empirical approach is a step in the right direction, however the analogous tests Sober uses fail for similar reasons as to the DA thought-experiments. Namely, they are not analogous enough. We then put forth the idea that a potential way to empirically test the DA itself would be to actually run entire-world simulations and forward-test future outcomes.

Lastly, Part IV discusses the broader implications of the DA. We suggest that even if the DA is correct, the consequences need not be as lugubrious as the name implies.

I. INTRODUCTION

1.0 INTRODUCTION TO THE DA

How long do we have left until the human race goes extinct? This is a question one might not normally expect to preoccupy formal epistemologists (as well as scientists, futurologists, and cosmologists), yet the Doomsday Argument [DA] has done precisely that for over thirty years. The DA is, in the simplest terms, an argument rooted in basic probability theory that aims to give a forecast about how many more humans will exist by factoring in the number of humans that have already existed.

Proponents of the DA suggest that it would be improbable for us to find ourselves somewhere near the start of the total existence of humankind. Once we consider our relatively low birth rank[1] and the notion that we are 'average' amongst observers,[2] an appropriate question to ask ourselves is, 'how likely is it that human beings will still be around millions of years from now?' Using basic probability theory, the DA shows that it would be rather unlikely for us to be in the beginning stages of our total existence. This notion, coupled with

---

[1] If we listed all humans that have ever existed, your birth rank would simply be your position in that sequence. You and I are roughly the 100th billion humans to ever exist (Population Reference Bureau).

[2] 'Average' in this sense means there is nothing particularly 'special' about us and our time in history with respect to our birth rank. We should therefore find ourselves taken from a random sample of all humans.

the rapid population growth rates of the last few centuries, leads the DA to give us a very pessimistic answer to the above question.

Although first informally conjured up by astrophysicist Brandon Carter,[3] the DA went on to be independently developed most notably by professor of astrophysical sciences Richard Gott III (1993) and philosopher John Leslie (1992a).[4] The DA still garners a rich academic interest today, and for almost every paper that gets written claiming to have 'debunked' the DA, another paper is subsequently published in reply, explaining why that objection is incorrect.[5] Nick Bostrom (2002a: 109) once quipped that DA objections "have a Phoenix-like tendency to keep re-emerging from their own ashes." Almost three decades after the DA was first formulated, it is our

---

[3] The phenomenon was subsequently dubbed the *Carter Catastrophe*. See Carter (1983).
[4] These are the two most popular expositions of the DA. It was also developed independently by at least half a dozen other individuals (including informally by Israeli entrepreneur Saar Wilf, as well as astrophysicist Stephen Barr) all roughly around the same time (Poundstone, 2019a; 24). See also Nielsen (1980) for a different formulation of the DA.
[5] As a result, the literature tends to be grouped into two camps: anti-doomsday, and anti-anti-doomsday. For seminal papers from the former camp, see Dieks (1992), Kopf et al. (1994), Tipler (1994), Tännsjö's (1997), Eckhardt (1993, 2013), Franceschi (2009), Goodman (1994), Smith (1998), Greenberg (1999), Lewis (2013), Sowers (2002), Caves (2000), Korb and Oliver (1998), Pisaturo (2009), and Hanson (1998). Replies to some of these objections and in defence of the DA can be found in Bostrom (2001a, 2007), Bostrom and Cirkovic (2003), Leslie (1992a, 1992b, 1993a, 1993b, 1996) and Gott (1994).

opinion that nobody has been able to conclusively show that the argument is incorrect, despite a plethora of attempts.[6]

Both Gott's and Leslie's DA rely crucially on the *Copernican Principle* [CP]. This is the notion that we are to be found in neither a central nor 'privileged' position in our universe.[7] Thus, if the human race goes on for millions of more years, we today should not find ourselves unusually early in the grand history of all of humanity. The DA borrows from this notion to estimate that our existence as a species will therefore not last much longer.

The DA also depends very much on the *Anthropic Principle* [AP]. This is the notion that observers (for example, us humans) could only find themselves in a universe that could produce such observers. Our position in space, as well as time, is in fact somewhat unusual because human life is only possible under rare circumstances. For example, our planet has all the right elements to facilitate our existence (there is enough carbon in the air for sustaining living organisms, we have access to water, and we are the ideal distance from the sun). Therefore, it is not a surprise that we find ourselves in this position because we could not have been anywhere else. The CP and AP form the basis behind *anthropic reasoning,* which just means that we must

---

[6] The PhilPapers section on the Doomsday Argument has in excess of 200 entries. A more contemporary overview of the DA literature can be found in Richmond (2006) and also in a recent book by William Poundstone (2019a).
[7] Copernicus proposed the radical theory that we are not at the centre of the universe with respect to our position *in space*. Here we apply the CP with regards to our position *in time*.

account for the fact that an observer is in a position to make an observation at all.

1.1 GOTT'S DA

The early DA literature of the 1990s tends to focus on either Gott's or Leslie's version of the DA. Gott's (1993) interpretation takes a broader form which he named the 'delta-t argument':

> Assuming that whatever we are measuring can be observed only in the interval between times t_{begin} and t_{end}, if there is nothing special about t_{now} we expect t_{now} to be randomly located in this interval.' (Gott, 1993: 315)

If we sketch a simple timeline of some event S:

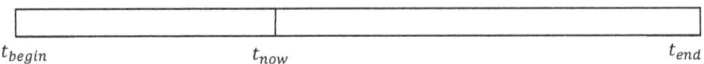

t_{begin} t_{now} t_{end}

Applying Gott's assumption of randomness (that we observe something of an unspecified duration at a *random point in its existence)*, he makes the following estimate:

$$t_{future} = (t_{end} - t_{now}) \approx t_{past} = (t_{now} - t_{begin})$$

Where t_{future} is an estimation of how long the series S will last.[8] Essentially, Gott (1993) is assuming that *past duration is a rough indicator of future duration*.[9] The series will go on for approximately as long as it did when we made the observation. Fifty percent of the time we will get an overestimate and fifty percent of the time we will get an underestimate. For instance, if we divide the timeline into quarters, we can construct a 50% confidence level:

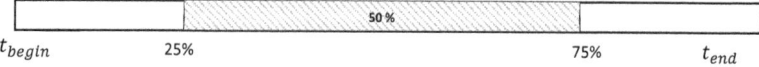

Suppose you encounter some event at the 25% point in its timeline. At that moment, the event's future will be 3 times the length of the past (as 75% is 3 times 25%). Alternatively, imagine that you are actually at the 75% point. Then the future will only be one-third of the past duration. Thus, the event's future duration will be between 1/3 and 3 times as long as it's past duration:

$$1/3 \cdot t_{past} < t_{future} < 3 \cdot t_{past}$$

---

[8] Strictly speaking, we are talking about expected values within this framework. Therefore, *on average*, we should expect these things to be equal. Gott is somewhat careless with explicitly stating this.
[9] This rather simplistic insight is very similar to the *Lindy's Law* – a phenomenon where the future life expectancy of something (usually a technology) is in proportion to its current age. See Taleb (2012: 159) and Eliazar (2017).

More generally if we want a prediction for some confidence level p:[10]

| $(1-p)/2$ | p | $(1-p)/2$ |

t_{begin} \hfill t_{end}

$$\frac{t_{now} - t_{begin}\left(\frac{1-p}{2}\right)}{p + \left(\frac{1-p}{2}\right)} < t_{end} - t_{now} < \frac{t_{now} - t_{begin}\left[p + \left(\frac{1-p}{2}\right)\right]}{\left(\frac{1-p}{2}\right)}$$

Substituting Gott's estimation that:

$$t_{future} = (t_{end} - t_{now}) \approx t_{past} = (t_{now} - t_{begin})$$

We obtain Gott's generalised DA formula:

$$t_{past} \cdot \frac{\left(\frac{1-p}{2}\right)}{p + \left(\frac{1-p}{2}\right)} < t_{future} < \frac{\left[p + \left(\frac{1-p}{2}\right)\right]}{\left(\frac{1-p}{2}\right)} \cdot t_{past}$$

Thus, for a 95% confidence level:[11]

---

[10] Where $0 < p < 1$.
[11] A 95% level is commonly used in the sciences. It indicates that the true value lies within the computed interval 95% of the time. Nonetheless, Gott is never particularly clear as to what makes 95% 'special' and why he chooses it specifically.

$$\frac{t_{now} - t_{begin}\left(\frac{1-0.95}{2}\right)}{0.95 + \left(\frac{1-0.95}{2}\right)} < t_{end} - t_{now} < \frac{t_{now} - t_{begin}\left[0.95 + \left(\frac{1-0.95}{2}\right)\right]}{\left(\frac{1-0.95}{2}\right)}$$

$$\frac{t_{now} - t_{begin}(0.025)}{0.975} < t_{end} - t_{now} < \frac{t_{now} - t_{begin}(0.975)}{(0.025)}$$

$$1/39 \cdot t_{past} < t_{future} < 39 \cdot t_{past}$$

Gott's inspiration for this idea famously came on a trip he made to the Berlin Wall:

> In 1969 I saw for the first time Stonehenge ($t_{past} \approx 3{,}868$ years) and the Berlin Wall ($t_{past} \approx 8$ years). Assuming that I am a random observer of the Wall, I expect to be located randomly in the time between t_{begin} and t_{end} (t_{end} occurs when the Wall is destroyed or there are no visitors left to observe it, whichever comes first). (Gott, 1993: 315)

By Gott's formulation and using the 95% level, the Berlin wall would cease to exist:

$$1/39 \cdot 8 \, years < t_{future} < 39 \cdot 8 \, years$$

$$0.21 \, years \, (\sim 11 \, weeks) < t_{future} < 312 \, years$$

As predicted, the Berlin Wall came down in 1989, and well within the 95% range.[12]

So how can we apply Gott's delta-t argument to our very own existence in time? Gott knew that at the time of publication, the first human emerged approximately 200,000 years ago. This gave him the following interval:

$$1/39 \cdot 200,000 \, years < t_{future} < 39 \cdot 200,000 \, years$$

$$5,128 \, years < t_{future} < 7.8 \, million \, years$$

This was the original way Gott interpreted the delta-t argument to form his DA, and the result is not particularly alarming. Many would probably comment that having 95% confidence that we will survive *at least* 5,000 more years incredibly optimistic, if not overly generous.

However, Gott (1994) later modified his approach to incorporate recent population booms and growth rates. He now asked us to instead imagine all humans to ever exist (past, current, and future) and to arrange them by their birth rank. It is not particularly clear what

---

[12] Note how even at the 50% confidence interval, Gott would have been correct.

exactly motivated Gott to make this change from thinking about how long humans have already existed to instead considering the number of births that have occurred. However, it highly likely stems from the exponential population growth we have witnessed over the last few centuries:

DIAGRAM 1: WORLD POPULATION GROWTH

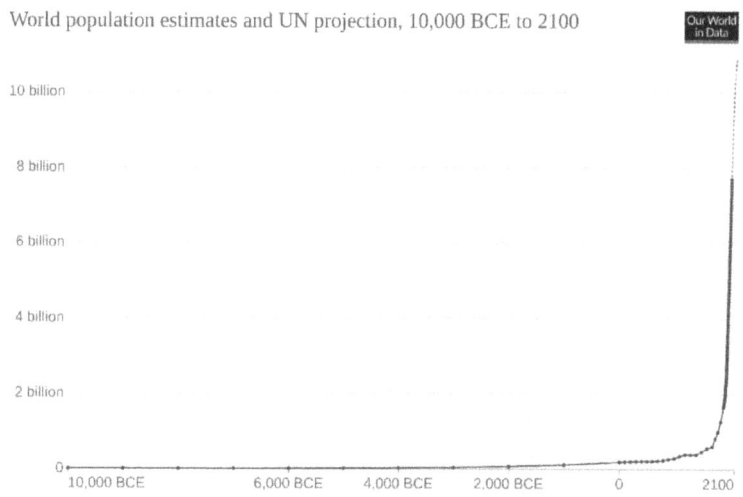

Nonetheless, this move by Gott strikes at the core of a key issue which we will explore in greater detail in 2.2. Namely, how exactly should one define the reference class.[13] As we will see in the following paragraph, using births instead of years gives a much smaller temporal confidence level because we are no longer applying the indifference principle to merely time.[14]

If we let N be the number of all observers who will ever exist and then apply the CP, we should find any single observer to be as equally likely to be at any point (taken from the total amount N). Importantly, we assume that $f = n/N$ (which is just our position across all humans expressed as a fraction) is spread evenly over the interval [0,1]. We will return to the validity of this assumption in section 3.1.

At Gott's time of writing, our birth rank was roughly 70 billion. What about the last human's birth rank? With 95% confidence, Gott projected that it would be somewhere between 1/39 and 39 times that number:

$$1/39 \cdot 70 \; billion < t_{future} < 39 \cdot 70 \; billion$$

---

[13] A temporary workable definition for *reference class* is simply the group to which a specific observer belongs.

[14] This issue is closely related to the Bertrand Paradox, where one can get wildly different predictions by sampling from different parameters using the same hypothesis. For his original critique of the principle of indifference see Bertrand (1889), and also Marinoff (1994). For the Bertrand Paradox within the context of the DA, see Vineberg (2011; 713-736), and Shackel (2008).

Thus, Gott predicted that the amount of people still to be born is somewhere between 1.8 billion and 2.7 trillion.[15] In order to translate potential future humans into years, we must take birth rates into account. When Gott was writing, there were roughly 150 million people born every year. At that rate, it would only take 12 years to gain an additional 1.8 billion (which has already occurred). The upper limit of the 95% confidence level would take 18 thousand years (in order to reach 2.7 trillion future births assuming a constant birth rate of 150 million a year).[16] By Gott's calculation, *doomsday will occur somewhere between twelve years and eighteen-thousand years* – the lower limit of which is incredibly disconcerting as we have already surpassed it.

Gott does not really elaborate on when exactly this method is applicable. While he encourages his readers to test out the delta-t argument on their current romantic relationships,[17] he says you may not apply it to estimate the marriage duration of newlyweds if you are at a wedding. The reasoning being that "you are at the wedding precisely to witness it's beginning" (Gott 1997: 39). Similarly, it would not work if estimating the total lifespan of our universe because "intelligent observers emerged only long after the Big Bang, and so

---

[15] This is for a 95% level. We will get very different results if we use 90% or 99%.

[16] This is a difficult assumption to justify. We will discuss the implications of a varying population growth later in 3.2.

[17] If you started dating someone a month ago, Gott predicts the relationship will last between 1/39th of a month (just under a day) and an upper limit of 39 months (slightly over 3 years) using the 95% confidence level.

witness only a subset of its timeline" (Gott, 1997:40). Both of these examples fail to satisfy the critical assumption that the observation was made at a time that is situated at a random point in the event's lifespan. Here we see the issue of assigning an appropriate reference class reappear. One could be forgiven for viewing Gott's dismissal of the wedding example and the age of our universe as appealing to an ad-hoc justification.

Gott (1997: 39) then claims that the delta-t argument is "most useful when examining the longevity of something, like the human race, for which there is no actuarial data available. We know only one human race. In predicting your lifetime, you can do better by using statistics on the life spans of people who have died." However, he does not provide any justification as to why a *singular* human race is important, and strangely ignores the fact that we *do* have empirical data for the longevity of other species. This foreshadows another issue which we will cover in detail in 2.2, a phenomenon known as the trumping issue.[18] Essentially, trumping can occur in instances where we in fact do have good empirical data and evidence that should outperform (or be a more valuable predictor than) any a-priori estimates we might get from DA-style calculations. As we shall later see, this may occur a lot more often than Gott would care to admit.

The delta-t argument appears to work well when tested against things like romantic relationships, Broadway plays, and the fall of landmark

---

[18] We borrow the coinage of this term from Northcott (2016).

walls, but it does stem from the crucial assumption that t_{now} is drawn at random (and in a uniform distribution) somewhere in the middle of t_{begin} and t_{end}. Moreover, we do indeed have actuarial data for all three of these things (a historical record of how long different relationships, Broadway plays, and walls have lasted in recent history). It is odd that nowhere in his analysis does Gott suggest that maybe looking at how long these things survived in the past could be seen as a better predictor. Gott (1997) seems to suggest that this method works best when we do not have any frequency data nor empirical evidence from which we could otherwise extract probabilities. When we do not have any empirical data related to the nature of the longevity of S, we are, according to Gott, entitled to assume that the distribution is flat. We will return to why this is an absurd assumption in our exposition of Sober's critique in 3.1.

Moreover, Gott's key assumption that we should treat our present situation as if it were sampled randomly from the entire timeline of S is very similar to invoking a kind of Principle of Indifference [PI] (Sober, 2003: 14). The PI is a rule which is often used to assign probabilities to epistemic outcomes.[19] It roughly states that when we lack relevant information, we should assign credence (belief) equally across all of the possible outcomes. This application of the PI in Gott's analysis has been criticised by Goodman (1994) for the reason

---

[19] An epistemic outcome is a statement about our knowledge of something.

that if we are ignorant of S's total longevity, we are also ignorant of the probability distribution.

Along with the aforementioned issues, Gott's argument has two further crucial shortcomings, one which will be addressed now, and the latter we save for 3.1. First, this method fails to consider the prior probabilities (which can be obtained empirically) of the hypothesis under consideration.[20] This is tangentially related to the aforementioned issue of trumping. For example, imagine you were a leading cast member on one of the Broadway plays Gott was estimating. With your insider knowledge, an estimate of when the final showing will take place is quite different from what a simple implementation of the delta-t argument would suggest.

Similarly, we have empirical predictions about risk factors that threaten our very survival (e.g., nuclear threats, pandemics, climate change models). These prior probabilities need to be considered when estimating future durations in situations where we have access to this kind of information.[21] The only time when Gott's formula would seem to work is when we restrict it to situations where in fact, we do

---

[20] The prior probability is the probability of an event before new evidence is obtained.
[21] Gott's original 1993 paper puzzled readers with its omission of prior probabilities. Poundstone (2019; 114) reached out to ask Gott why he ignored Bayes, to which Gott responded, "'*Bayesians*... I didn't put any Bayesian statistics in this paper because I didn't want to muddy the waters... Because Bayesian people will argue about their priors, endlessly. *I had a falsifiable hypothesis.*"

lack any other relevant information. This is obviously not the case when approximating the future longevity of our species as we have a plethora of additional information that would be relevant. Again, we run into the issue of trumping, albeit in a slightly different setting. For these reasons, we think that Gott's DA is incorrect. The final nail in the coffin concerns the underlying sampling process which we will cover shortly.

Due to these rather serious objections, the academic DA literature of the 21st century tends to focus almost exclusively on Leslie's version of the DA. Four notable exceptions are Monton & Roush (2001), Pisaturo (2009: 2011), Monton & Kierland (2006), and Sober (2003), whose critique we will explore in due course. Apart from these publications, most analyses of Gott's DA have been relegated to an occasional appearance in non-academic journals and an informal discussion of the DA in popular news media.[22] This has coincided with an increased scholarly interest in Leslie's version of the DA.

1.2 LESLIE'S DA

Leslie presents a less formal, however philosophically more captivating interpretation of the DA, and one that actually does take the empirical priors into account. Leslie's account is presented as a

---

[22] For example, see recent articles in the Wall Street Journal (Poundstone, 2019b) and Vox (Poundstone, 2019c).

mixture of analogies and hypotheticals, which makes it difficult to summarize as such. Perhaps the best way to formally introduce Leslie's DA (and in the spirit of this dissertation title) is with a thought-experiment. Consider the following three-stage hypothetical:[23]

Step I: Imagine there are 1000 stalls, 70% are coloured blue, 30% are coloured red. You happen to find yourself inside one of these stalls. If somebody would ask you, 'how likely is it that you were put inside a blue stall?' How should you reason? Well, absent of any additional information, you should think that there is a 70% chance that you were put in a blue stall because 70% of all stalls are blue. This is what Bostrom (1999, 2000, 2002a) calls the Self-Sampling Assumption:

> *Self-Sampling Assumption [SSA]:* An observer should reason as if he or she were a random sample drawn from the set of all observers.

At the outset, this seems a rather straightforward line of reasoning (and it follows naturally from the CP and AP stated in the introduction), but as we shall see when covering the Shooting Room Paradox, the SSA can be fairly controversial. Nonetheless, in this example, nearly everyone would think that you were placed in a blue stall with 70% probability. In general, we can say that one's credence

---

[23] This thought-experiment is reworked from an exposition given by Bostrom (2004).

of having some property *P* should correspond to the fraction of observers who also share that property *P*.

Step II: Start over and now pretend that we have 1000 stalls numbered and marked 1-1000 on the outside. We flip a coin and if it comes up heads then observers will only be put into stalls 1 to 10. If tails, then observers will be put in all 1000 stalls. You happen to find yourself in a stall and are asked if you are in the 1000-people or 10-people scenario. You should think that either scenario was 50% likely as the result was decided by a coin flip, and you do not have any additional information.

Suppose now that you walk outside the stall, and you observe the number six drawn on the door. Do you still hold on to your previous estimates?[24] No, you would have good reason for revising as it would be much less probable you are in the 1000-stall scenario than in the 10-stall scenario.[25] To calculate just how much less probable, we turn to Bayes. Specifically, Bayesian reasoning tells us that the probability of some evidence *e* that a certain hypothesis *h* is correct will decrease or increase in proportion to any additional likelihood that you would

---

[24] To see why, just remind yourself that:
    i)   The odds of being in number 6 if there are 10 stalls is 1/10.
    ii)   The odds of being in number 6 if there are 1000 stalls is 1/1000.

[25] If this is still intuitively difficult to grasp, substitute an even larger number for 1000 such as 1 million. Would it not be it extremely 'odd' that you are amongst the first 10 if there was a possibility out of 1 million?

have gotten if that hypothesis was correct. Consider the following notation:

$P(h,e)$ — the probability that h is correct given evidence e.

$P(h)$ — the prior' probability.[26]

$P(e,h)$ — the probability of getting evidence e if hypothesis h were correct.

$P(e)$ — the prior probability of evidence e

Bayes Rule [BR] gives us the following relationship:

$$P(h,e) = \frac{[P(h)P(e,h)]}{P(h)P(e,h) + P(not\ h)P(e, not\ h)} = \frac{[P(h)P(e,h)]}{P(e)}$$

Using the above example of 10 or 1000 stalls, how likely is it that you were placed in stalls 1-10? There was a 0.5 chance that the coin was heads, and you were in stalls 1-10, so our prior probability is *P(h)* = 0.5. We need to find $P(h,e)$ given that

$P(h) = 50\%$,

$P(e,h) = 1/10$

---

[26] Prior here simply means the probability that *h* is correct *before* we take evidence *e* into account.

$P(\text{not } h) = 50\%$

$P(e) = 1/1000$

Hence,

$$P(h, e) = \frac{50\% \times \frac{1}{10}}{50\% \times \frac{1}{10} + 50\% \cdot \frac{1}{1000}} = \frac{0.05}{0.05 + 0.0005} = 0.990 \ldots \approx 99\%$$

Whereas before we were 50% confident that the coin landed tails, after seeing the number 6 on our door we are now almost certain (99%) that it fell heads, which must mean the 1-10 stall scenario. After finding out that you are in stall number 6, your credence that the coin landed heads should be 99%.

Now we need to apply these insights to our situation here on earth.

Step III: Imagine that we have two rival hypotheses:

DoomSoon – Humankind will die out within 100 years. A total of 200billion ($200 \cdot 10^9$) humans will exist in total.

DoomLater – Humankind will survive the next 300billion years. A total of 200trillion ($200 \cdot 10^{12}$) humans will exist in total.

Although this exposition greatly simplifies our reality (and these concerns will be addressed shortly), for this demonstration it does not really matter if the numbers are precise and nor do we need to limit the possibilities to only two options.[27] DoomSoon corresponds to the 10-stall scenario in Step II. DoomLater corresponds to the 1000-stall scenario.

Suppose also that we can estimate some prior probabilities for these outcomes. These will be calculated from our best empirical data regarding the existential threats facing our survival (such as nuclear war, viruses, asteroids, dangers of nanotechnology and so on). Now assume that based on all these considerations, you calculate that there is a 5% probability of DoomSoon and thus a 95% probability of DoomLater (the specific percentage amounts are not important in order to illustrate how the argument works). These estimates form our priors.

Instead of numbers written outside the stall, we have our 'birth rank'. This is your position amongst all humans in the human race. For example, our birth rank is somewhere around 100 billion. That is how many people existed before we were born. Just as finding out that you were in stall 6 increased your probability of the coin having landed on heads, finding out that you are birth rank 100 billion should give

---

[27] See Bostrom (1999) for an explanation that shows how the math remains much the same even if we have more than two possible outcomes.

you reason to think DoomSoon is much more likely now. To calculate exactly how much more likely, we again turn to Bayes:

$P(h) = P(DoomSoon) = 0.05$

$P(e, h) = 1/200 \; billion$

$P(not\; h) = P(DoomLater) = 0.95$

$P(e) = 1/200,000 \; trillion$

Hence,

$$P(DoomSoon, 100bn) = \frac{0.05 \left[\frac{1}{200 \cdot 10^9}\right]}{0.05 \left[\frac{1}{200 \cdot 10^9}\right] + 0.95 \left[\frac{1}{200 \cdot 10^{12}}\right]} = 0.9814 \ldots \approx 98\%$$

Whereas previously we were only 5% sure that humankind would die out in the next century, the posterior probability of DoomSoon now becomes almost certain. After conditionalizing upon the fact that our birth rank is 100 billion, we ought to seriously consider ruling out the possibility of DoomLater.[28]

This is the basic idea behind Leslie's DA. If we had trillions of humans born throughout all of history, then a human located at the turn of the 21st century would be considered exceptionally early. We

---

[28] A more generalized version of this is given in Bostrom's (2002b) Amnesia Chamber thought-experiment. There, he shows why it must follow from Bayesian principles that when presented with the two options 'MANY' future individuals or 'FEW' future individuals, learning *e* (which is the new piece of evidence that is your birth rank), will bring down the likelihood of MANY and increases that of FEW.

can express the updated probability of DoomSoon given our birth rank more generally and get Leslie's general DA formula:

$$P(DoomSoon, Br) = \frac{P(DoomSoon) \cdot P(Br, DoomSoon)}{P(DoomSoon) \cdot P(Br, DoomSoon) + P(DoomLater)P(Br, DoomLater)}$$

Where,

$Br = birthrank$

$DS_n = $ total number of humans that live in DoomSoon scenario

$DL_n = $ total number of humans that live in DoomLater scenario

As the CP requires us to treat any person as being equally likely to be located at any position across the total population:

$P(Br, DoomSoon) = 1/DS_n$

$P(Br, DoomLater) = 1/DL_n$

Which finally gives us

$$P(DoomSoon, Br) = \frac{P(DoomSoon) \cdot \frac{1}{DS_n}}{P(DoomSoon) \cdot \frac{1}{DS_n} + P(DoomLater) \cdot \frac{1}{DL_n}}$$

In Leslie's version of the DA, this is how the calculations are done. It should be noted though that the above equation is *not* Leslie's DA.

Rather, this equation is derived from specific assumptions that Leslie's DA tries to defend. Notice also that your specific birth rank figure does not actually feature in the calculation itself, we are only conditionalizing on the fact that you exist now.[29]

It is worth mentioning how varying our initial assumptions can change the outcome drastically. For example, suppose we thought that *P(DoomSoon)* and *P(DoomLater)* were roughly equal (both fifty percent). On this assumption, the probability of DoomSoon *after* conditionalizing on our birth rank becomes nearly 99.9%.[30]

Alternatively, Leslie (2010: 450) suggests that maybe the human race could survive long enough to achieve interstellar travel and colonize other planets, thereby causing the population of humans to grow to 50 quadrillion. On this assumption, even if we assign $P(DoomSoon)$ a very small chance (e.g., 1%), conditionalizing on the idea that we are part of the first 100 billion humans would give a probability of 99.9% for $P(DoomSoon, 100bn)$.[31]

Another consequence is that the more humans we think will exist in the future, the larger the probability of DoomSoon (all other things

---

[29] The intuition here is clearer if you think back to the stalls. Did it really matter that it was '6' on the door and not '1' or '5' or '9'? No, and the Bayesian calculation would be the same in all of those cases.

[30] $P(DoomSoon, 100bn) = 0.5\left(\frac{1}{200 \cdot 10^9}\right) / \left[0.5\left(\frac{1}{200 \cdot 10^9}\right) + 0.5\left(\frac{1}{200 \cdot 10^{12}}\right)\right] = 0.99900 \ldots \approx 99.9\%$

[31] $P(DoomSoon, 100bn) = 0.01\left(\frac{1}{200 \cdot 10^9}\right) / \left[0.01\left(\frac{1}{200 \cdot 10^9}\right) + 0.99\left(\frac{1}{5 \cdot 10^{16}}\right)\right] = 0.99960 \ldots \approx 99.9\%$

being equal).³² This can be seen by a simple inspection of the previous equation.

Leslie's DA differs from Gott's in at least three significant ways. First, Leslie does not specify a range or a prediction, as per Gott. Instead, he merely suggests that a probability shift *must* occur once we consider our place amongst all observers who will ever exist when comparing DoomSoon and DoomLater.

Second, Leslie's DA is indifferent to *how long* we, as a species, have already existed (as this does not factor into the calculations). Instead, the relevant piece of information is that *I am an observer existing today*. Past duration is irrelevant in Leslie's case. All else being equal however, Leslie's calculation does take your birth rank into account because we are conditionalizing upon it (even though the exact value doesn't get involved in the calculation per se).

Third, whereas Gott places the current moment roughly at the halfway point of our species total duration, Leslie's DA suggests that we are closer to the end. Leslie's DA is interesting in that it relies on basic

---

³² How population pressures may magnify potential causes of disaster is examined in Leslie (1994: 35). A greater *P(DoomSoon)* would mean that it is less likely that the total amount of humans will be a large figure associated with DoomLater. If the only two options are a low or a high figure, then the low figure is *relatively more likely* if the high figure becomes abnormally large. Leslie's calculation only concerns *relative* chances of the two figures on the assumption that those are the only two possibilities. It does not, however, help calculate the expected number of observers across the full range of outcomes.

probability theory yet, if correct, could have enormous implications. The fact that these probability shifts that occur could be infinitesimally small does not in any way undermine Leslie's argument.

II. DOOMSDAY ARGUMENT THOUGHT-EXPERIMENTS

2.0 INTRODUCTION TO DA THOUGHT-EXPERIMENTS

The use of thought-experiments in philosophy is not uncommon as they enable us to illustrate complex phenomena by invoking our sense of imagination. There has been a tendency in the literature for philosophers to try to debunk the DA by introducing thought-experiments that show why it must be wrong. This paper will not address the more general philosophical issues associated with thought-experiments.[33] Instead, we will argue that the use of thought-experiments within the context of the DA has serious limitations. It should also be noted that the motivations for DA thought-experiments have more in common with the sciences than that of philosophy, mainly because we are using them to gain an insight on an experiment which, unfortunately, is physically impossible to conduct.

There are at least 40 different thought-experiments which have appeared in DA literature since the argument's inception.[34] Some, like Adam and Eve (Bostrom, 2001a) or The Emeralds (Leslie, 1996;

---

[33] See Kuhne (2007); Sorensen (1992), and the SEP entry on thought-experiments.
[34] As no current anthology of DA thought-experiments exists, we have assembled a compendium (though perhaps not exhaustive) in Appendix A.

220) have been devised specifically to tackle some aspect of the DA. In fact, most of the thought-experiments in DA literature are of this variety. Others, however, like the Sleeping Beauty Paradox (Zuboff, 1990), or the Monty Hall Problem (Selvin, 1975; vos Savant, 1997), were developed independently of the DA but have proved very useful in their application to the argument. Rather than strive for completeness, we have selected three prominent DA thought-experiments for critical examination.

The three examples that will be covered are Leslie's (1989a) Urns, Eckhardt's (1997) Shooting Room, and Northcott's (2016) Asteroids. Our reason for choosing this particular trio is that they span three decades of DA literature and as a result neatly encapsulate how thinking about the DA has evolved. More importantly, each one presents a unique obstacle, or 'conceptual challenge' for the DA. In our chosen three, we will be exploring, respectively, the issues of reference class, trumping, and determinism. All three concepts are also intricately interwoven, as we will see in 3.3. Our aim here is to show that while these thought-experiments are useful tools for making us think more clearly about what the DA implies, any attempt to *refute* the DA via thought-experiment will ultimately fail because of the unique nature of the argument. If we want to disprove the DA, we will need to go beyond thought-experiments and get our hands dirty with empirical evidence, which is what we attempt to do in Part III.

2.1 LESLIE'S URNS

The Urns thought-experiment was first introduced in a paper by Leslie (1989a) and then again more formally in Leslie's (1996) locus classicus *The End of the World*. It has subsequently become the gold-standard for DA analogies. Due to its simplicity as an intuition pump, it is perhaps the most popular DA thought-experiment, and still commonly recited in philosophy classrooms as a means of introducing students to the argument:

> Suppose there is a 98 percent probability that a lottery urn with my name in it contains one thousand names, and a 2 percent probability that it contains just ten... What if I next find that mine is among the first three names drawn from the urn? Bayesian calculations give me a new estimate...Calculations on similar lines can suggest that the risk that the human race will end soon has been regularly underestimated, perhaps very severely. (Leslie, 1996: 201)

The updated estimate that the ball came from the 10-ball urn is now roughly two-thirds.[35] Note how this urns analogy is not too dissimilar

---

[35] $P(10\ ball, among\ first\ 3) = 0.02(0.3)/[0.02(0.3) + 0.002(0.98)] = 0.671... \approx 67\%$

from the stalls thought-experiment already discussed in the introduction.[36] The question we must now ask ourselves is whether pulling names from an urn is *analogous enough* to the case of us being born into this world?[37]

While it is obviously true that we are not balls in an urn nor persons inside stalls, analogies of this sort are used often in statistics classes to illustrate probabilistic arguments. Objectors will claim that the Urns seem like an armchair philosopher's attempt to predict the apocalypse, however thought-experiments of a similar nature are part and parcel to certain other areas of philosophy. It is not enough to just say we are not balls in an urn. The onus falls on the objector to identify exactly what is different about the urns case and our reality.

2.2 CONCEPTUAL ISSUE IN LESLIE'S URNS: REFERENCE CLASS

While Leslie's Urns serve as a great thought-experiment to introduce the basic framework of the DA to the reader, they are also

---

[36] We turn the focus from stalls back to urns for two reasons. First, to remain faithful to the original way in which Leslie presented the thought-experiment. Second, in our experience, *stalls* are a better metaphor when we want to explain the basic intuition of the DA. Using *balls in urns* is better when we want to focus on what we discuss in this section, namely *reference class*. Structurally, however, there are no substantive differences between the two expositions.

[37] See Leslie (1996: Ch.11) for a systematic discussion of common objections to this thought-experiment and why they are incorrect.

instrumental in highlighting the issue of choosing a suitable *reference class*.[38] It is our view that there are actually two *types* of reference class issues involved in the DA, and to help better understand our conceptualization of each, consider the following sketch of human evolution:

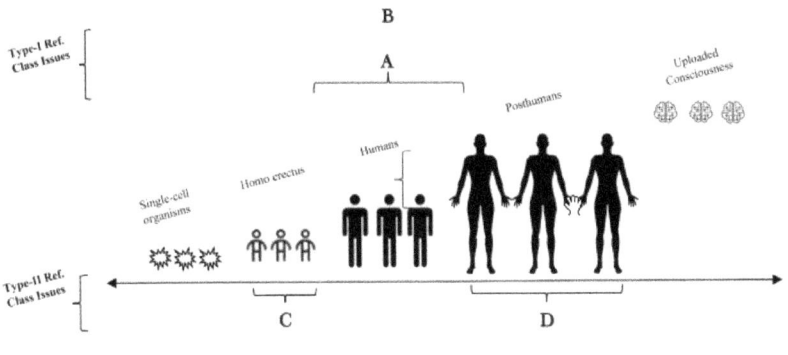

For simplicity, let us assume that we evolve in five 'stages' from single-cell sentient organisms to *homo erectus* to 'humans' to then 'posthumans'[39] and finally fully digitalized and uploaded consciousnesses. We acknowledge that this is a gross oversimplification of how evolution works, but it is necessary here

---

[38] For a more general explanation on probabilistic issues with assigning reference class, see Hajek (2007).

[39] Posthumans is a term borrowed from futurology literature that means more technologically evolved humans, to a point where we are significantly distinguishable from today's humans in intelligence.

for us to make our point. Type-I reference class issues are concerned with questions such as *where do we draw the line between human and homo erectus*? Or human and posthuman? Do we make our DA calculations over region A or over region B? How do we decide where to make the demarcation? This is crucial because, as we shall see, we will get wildly different results depending on who we choose to include.

Type-II reference class issues are concerned with the following: does DA style reasoning apply to class C *and also* class D? Essentially, which *things* does DA style reasoning concern? We use the vague term 'things' because Type-II issues need not be limited to just classes of species, as we already saw when covering Gott's DA and how it does apply to Broadway plays and relationships but does not apply to the lifetime of our universe (or relationships at the start of a wedding ceremony).[40]

Although Type-I and Type-II reference class issues are clearly intertwined, the DA literature tends to focus almost exclusively on Type-I issues. This is understandable for two reasons. First, because

---

[40] A more topical application of Gott's delta-t argument can be found in an interesting paper by Costas et. al (2020) titled *Groundbreaking predictions about Covid-19: a novel mathematical approach never used in epidemiology*. Interestingly, the authors (publishing in July of 2020) predicted that "according to the Doomsday principle, the Covid-19 pandemic should end between September 2021 and October 2022". As this dissertation is being submitted before September 2022, we cannot verify if their prediction was correct.

philosophers and cosmologists (Gott aside) working in this area are primarily concerned with the big question of 'when will *we* go extinct?' They can be forgiven for not being particularly concerned about predicting the fall of famous walls or when relationships will end. The second reason, as we shall now see, is because Leslie's exposition of the DA only really concerns Type-I reference class issues – who is in *our* reference class? In Leslie's writing, he never makes the claim that his DA can be applied to other things or species. He is solely focused on the issue of human extinction.

Nonetheless, we believe it is a mistake to ignore Type-II reference class issues as they may potentially provide the key to disproving the DA. If we can show that the DA fails for other analogous *things*, we may have good reason to reject the DA when applied to our longevity. This is exactly the direction that Sober (2003) takes, as we will see in Part III.

For now, let us focus on Type-I reference class issues by returning to Leslie's urn analogy. Consider the following objection: *the names in the urn are a well-defined set. How do we deal with our Homo erectus predecessors or the possibility of future posthumans?* This gets at the heart of Type-I reference class issues, and it has generated a lively debate in the DA literature (see Bostrom, 1999; Olum, 2002; Franceschi, 2009; Simpson, 2016). In our earlier discussion of the SSA, we concluded that we should reason *as if* we were randomly selected from a random sample of observers (specifically, observers in our reference class):

$Cr(I\ have\ P\ |\ A\ fraction\ of\ x\ of\ all\ observers\ within\ my\ reference\ class\ have\ X) = x$

The question still remains – who is in our reference class? The standard response in early DA literature is rather vague and usually merely states 'humans' without much elaboration. However, who exactly classifies as 'human' has been actively contested on philosophical (as well as rational) grounds. There was a 'fuzzy' period where homo erectus were not fully homo sapiens and homo sapiens were not fully evolved homo erectus. How much does this change our DA calculations? It turns out, rather significantly.

Leslie's proposed solution here is rather straightforward. We can simply choose the reference class to be whatever we want it to be. Once this delineation has been made, the DA still works. Leslie (1996: 255) illustrates this by slightly altering his thought-experiment and having the urn balls painted in an assortment of colors (while still keeping them numbered). Imagine you are now tasked with guessing how many 'green' balls there are left in the urn. Of course, 'green' is somewhat of an ambiguous concept. When does green become aqua, or green-blue, or lime-green? This vagueness corresponds exactly to the vagueness one encounters when delineating who exactly is an observer in the DA. Leslie's solution is simple: just choose whichever way you wish to define the reference class. Imagine you want to know the amount of balls that are green-to-bluish. Then all that would be required would be to separate the random sample into green-to-bluish and *not* green-to-bluish. After having made this reference class categorization, the calculations are the same as before. Bostrom (2002a; 106) echoes this sentiment, noting that, "the Bayesian apparatus is neutral as to how you define hypotheses; there is no right or wrong way, just different questions you might be interested in asking."

However, something still does not seem quite right when indiscriminately applying this reasoning. As Leslie (2010) himself later realized, for the purposes of his argument he needed to place *some* kind of restriction of which reference classes are permitted, lest we risk having classes that are either too narrow or far too wide. The consequences here can have rather radical implications. For example, if we indeed include all previous organisms going way back to single-cell, then all of a sudden the DA calculations look rather innocuous. For many calculations of DoomSoon and DoomLater, we would now have a shift *in favor* of DoomLater simply by adjusting our reference class. This would, in effect, defeat the whole purpose of the argument.

We think one possible resolution to the issue is to make *consciousness* the primary discriminator (or at least part of it). Consider Bostrom's (2002a: 162) concept of *observer moments,* in what he calls his Strong Self-Sampling Assumption:

> SSSA: each observer-moment should reason as if it were randomly selected from the class of all observer-moments in its reference class.

We still need to ask ourselves what exactly is an 'observer-moment'? It can be safely assumed that in order to register an 'observer-moment', one must be somehow part of a conscious agent. There is certainly an aspect that cognition is a requirement for one to experience an observer-moment. This seems, to us, a reasonable demarcation of reference class. If we decide that is what we are concerned with, then the DA still applies. Leslie (1996: 260) had a similar idea:

> The moral could seem to be that one's reference class might be made more or less what one liked for doomsday argument purposes. What if one wanted to count our much-modified descendants, perhaps with three arms or with godlike intelligence, as 'genuinely human'? There would be nothing wrong with this. Yet if we were instead interested in the future only of two-armed humans, or of humans with intelligence much like that of humans today, then there would be nothing wrong in refusing to count any others.

While we mostly agree with the above rationale, we think Leslie overlooks two crucial pitfalls. The first is related to Type-II reference class issues, the second to Type-I. Consider the following true statements:

1. I am blue-eyed. I am located randomly amongst all blue-eyed people that will ever exist

2. I am Russian. I am located randomly amongst all Russians that will ever exist.

3. I am a Birkbeck MRes Philosophy student. I am located randomly amongst all Birkbeck MRes Philosophy students that will ever exist.

Lastly, I am also a human observer, and, according to the DA, I should find myself located randomly amongst the set of all human observers to ever exist. Performing DA style calculations for 1-3 will yield different, and most likely contradictory results.[41] Leslie would retort here that this is not an issue – as long as you adjust your priors accordingly:[42]

> No inappropriately frightening doomsday argument will result from narrowing your reference class . . .

---

[41] Although not necessarily. Perhaps blue eyes will evolve away before doomsday? There is also no reason to believe that Birkbeck (or at least the MRes course) will still be around for the end of times.

[42] Leslie attributes this idea to a suggestion made to him by Brandon Carter.

provided you adjust your prior probabilities accordingly. Imagine that you'd been born knowing all about Bayesian calculations and about human history. The prior probability of the human race ending in the very week you were born ought presumably to have struck you as extremely tiny. And that's quite enough to allow us to say the following: that although, if the human race had been going to last for another century, people born in the week in question would have been exceptionally early in the class of those-born-either-in-that-week-or-in-the following-century, this would have been a poor reason for you to expect the race to end in that week, instead of lasting for another century. (Leslie 1996: 262)

But how far can we keep pushing this narrowing of reference class until we crash into absurdity? Bostrom (2002a: 105) heeds the following warning: "alas, it is a vain hope that the prior will cancel out the distortions of a gerrymandered reference class." Indeed, if we consider my three statements above, which 'priors' should we choose in each calculation? The worry is that with any given priors, we will get varying and contradictory estimates. It is very much contingent on the reference class used. While it is trivially correct that when using any arbitrarily chosen reference class, we can always cherry-pick some numbers in a manner that when plugged into Bayes formula (together with the priors associated with that reference class), we will get posterior probability functions that are consistent with each other. However, as these values were arbitrarily chosen, they will not *really* represent the prior probabilities.

The second issue stems from the latter part of Leslie's quote. We are not only interested in the longevity of the "two-armed humans" of today. It is understandable that at the time of Leslie's writing in the early 90s, most would not have seriously entertained the idea that we could create machines with human-level cognition within the next century. Today, however, a significant portion of modern futurologists believe that we are in fact mere decades away from doing so.[43] What happens if we evolve into posthumans before doomsday? Or if we can download our conscious minds to live *in silico* for eternity? Within the framework of the DA, the 'human race' as we know it could reach 'extinction' imminently, however this conclusion is not as pessimistic as previously imagined. We will return to this idea in more detail in Part IV.

In Leslie's Urns analogy, the DA works because we have a well-defined discrete reference class of objects. When applying the DA to human births, we need to choose something that is also sequential and distributed in such a way that any position on that sequence is just as likely to contain a human observer. Even DA-proponent Nick

---

[43] The literature on mind-uploading consciousness and transhumanism is rich and vast. As a good starting point, we recommend the following: Sandberg & Bostrom (2008); O'Connell (2018); Hofkirchner & Kreowski (2021).

Bostrom (1999: 541) himself once conceded that, "in my opinion, the problem of the reference class is still unsolved, and it is a serious one for the doomsayer."[44]

Although we do agree that the reference class poses an issue, we do not think it is a fatal blow to the DA. Leslie's coloured urns analogy gives us a potential solution to the problem, however the question of how to properly define reference class still remains. Ultimately, we believe, it would be incorrect to view the reference class issue as a full and concrete refutation of the DA.

2.3 ECKHARTD'S SHOOTING ROOM[45]

Here is the Shooting Room [SR] thought-experiment as described by Leslie (1996: 251):

> Imagine that the Devil creates people in a room, in a batch or several batches. Successive batches would be of 10 people, then 100, then 1,000, then 10,000, then 100,000, and so forth: each batch ten times as large as its

---

[44] He does appear more impartial to the issue in later writings; cf. Bostrom (2003, 2007).

[45] We attribute this thought-experiment to William Eckhardt, despite the fact that it was Leslie (in personal correspondence with David Lewis) who actually devised the basic set-up. Eckhardt's name, however, is most synonymous with this experiment as he went on to modify it and provide a 'solution' in two subsequent papers.

predecessor. You find yourself in one such batch. At every stage in the experiment, whether there will be any later batches is decided by the Devil's two dice, thrown together. The people in any batch watch the dice as they fall. So long as they don't fall double-six, the batch will exit from the room safely. If they fall double-six everybody in the batch is to be shot, the experiment then ending. How confident should you be of leaving the room safely?

If you find yourself inside the SR, you will most likely reason that there is a 1-in-36 or a 2.78% chance of getting shot, which is just the odds of a double six dice roll (1/6 x 1/6). However, it is a fact that 90 percent of everyone who enters the room will perish. Should your odds of survival therefore not be a mere 1-in-10?

From the 2nd round onwards, 90% of everyone who has entered the room will die.[46] Upon first hearing about the SR, most will probably be tempted to side with the 1/36 explanation. Note, however, that any insurance company paying out life policies would lose money if it accepted the 1-in-36 odds. For it to be profitable, the pay-out must be contingent on the fact that 90% of all those who enter the room will die.

---

[46] The exception here is if double sixes land on the first roll. In that case, and that case only, the first person is killed and the game ends, so the death rate is 100%.

We can further increase the absurdity in the paradox. Consider Bob and his grandmother, Sally. One day Bob is called to the SR. He messages Sally telling her not to be worried. He figures his odds of getting shot are a mere 1-in-36. Later that day Sally, being of old age, loses her cell-phone on the underground and cannot receive any more messages. She gets home and sees on the news: 'BREAKING; Shooting Room Slaughter Strikes Again'. To poor Sally, the rational belief now is that Bob was amongst the unlucky 90%.

So, what does the SR have to do with the DA? Well, we could think of the thought-experiment as almost a cartoonish allegory for our own reality. On any given day, our risk of extinction is pretty small. However, we know that there will come a day when our luck stops. We know that our population has been increasing exponentially over the last few centuries. Of course, that is not to say it will keep growing exponentially.[47] However, it is estimated that roughly 10% of all people that ever existed are alive today (thus, not quite the 90% as depicted in this analogy). The SR is nonetheless a useful intuition pump for our situation here on earth.[48]

---

[47] In fact, in 2022 many countries are already facing a declining growth rate: "the global growth rate in absolute numbers accelerated to a peak of 92.9 million in 1988 but has declined to 81.3 million in 2020. Long-term projections indicate that the growth rate of the human population of this planet will continue to decline, and that by the end of the 21st century, it will reach zero." [UN World Population Prospects]

[48] Of course, some of the finer details are significantly different. For example, the values in the paradox are not 'realistic' in the sense that, on average, one would need to roll two dice 36 times before they got a double six. The room by the 36th round would have 10^{34} people, which far exceeds

2.4 CONCEPTUAL ISSUE IN ECKHARDT'S SHOOTING ROOM: DETERMINISM

The SR is a peculiar paradox as it seems to hinge on the result of a dice throw, which is amongst the simplest of exercises in probability. How could there be anything mysterious about something that has been understood for hundreds of years? [49] Let's analyse how the situations differ based on the observer's perspective. For Sally, the SR is somewhat of a black-box as she can only interpret how it operates from the 'outside'. Sally knows that 9 out of 10 who enter the room will not leave alive. She uses that information to form her credence. Bob, however, focuses on the dice. He is aware that his destiny depends solely on the next roll. Using probability, he estimates a 0.972% chance of survival. In this instance, Bob's credence is more 'correct' because Bob knows about both sides of the experiment. Sally is unaware about the more detailed workings of the SR. She cannot factor in the roll of the dice, and only fixates on the '90% are dead' headline. In that case, her answer is also justifiable given that she has incomplete information.

The experiment is designed to illustrate the possibility of one event having two 'correct' probabilities. It shows two sides to understanding risks in the DA. Bob is akin to someone who has an

---

any realistic estimates of future population size. We would surpass the current world population already by the 12th roll.

[49] David Lewis has been quoted as saying that the SR is "a good, hard paradox" (Poundstone, 2019; 234).

understanding of all of the empirical and existential risks that plague human survival, and he is able to obtain correct probabilities for them. Sally is more like someone who uses strict DA reasoning (Leslie's, more so than Gott's). In that sense, her understanding of someone in the SR is akin to Leslie's understanding of us being randomly sampled from a rapidly increasing population.

For Leslie (1996: 251-255) however, the SR paradox boils down to *determinism*. Throughout his exposition, Leslie repeatedly stresses that if the world is fundamentally indeterministic (as many quantum physicists believe) then the DA is substantially weakened. Indeterminism would imply that there was no suitable 'firm fact of the matter' (in theory available to anyone who knew the current situation and the physical laws in sufficient detail) regarding how many humans are yet to come into existence before doomsday occurs.

When she sees the news headline, Sally accepts that Bob had met his fate. At that very moment we could print out a list of names of everyone who was ever inside the SR, and 90% of those would be killed. Although Sally doesn't have access to this list, she has good reason for believing that Bob's name is not amongst the 10% of survivors. For Bob, standing within the walls of the SR, his destiny is still undecided. The pivotal roll of the dice is yet to occur, so the outcome for him is somewhat unpredictable. At that very moment in time, it is impossible for anyone inside the SR to compile a full list of names of future and present occupants. Bob has more reason to fixate on the outcome of the roll rather than believe that he is randomly plucked from a long list of names.

Similarly, the DA asks us to imagine drawing a name from all past, current, and future observers. In principle, this list could potentially already exist, *if the world is predetermined.* This is what Leslie believes to be the Achilles heel of the DA. Coming back to the Urns analogy, consider how many balls remain in an urn after your number has been drawn. If the answer is undefined, this would thwart Leslie's efforts to re-estimate the risk of DoomSoon. The DA is persuasive *only if* hard determinism is a feature of our reality. If not, Leslie believes that the DA is seriously undermined.

To cement his point, Leslie (1996: 254) gives an altered version of the SR where everything is deterministic:

> Suppose the Devil had decided to let each batch leave the Shooting Room safely until successive digits of pi (the ratio of a circle's circumference to its diameter, whose first digits are 3.14159...) yielded a double—six. After the seven millionth digit (a point which the Devil, no mathematician, chose without knowing what this digit would be and which digits would come next), successive digits were to be taken two by two: just two new digits whenever a new batch was in the room. You find yourself in the room. The Devil's computer is

calculating what the next two digits will be. Should you expect them to be 6 and 6?

In this case, we definitely should. Someone who finds himself in the room in this set-up cannot say that their fate has been sealed by factors that are deterministic, as the numbers in pi are wholly indeterministic. The key point being made here is that, given determinism, if all those who are in the room expected death, *the majority of people would be right.*

For William Eckhardt (1993: 484) however, "the issue of determinism is a red herring". In his paper *A Shooting Room View of Doomsday*, he presents a very convincing argument as to why Leslie is mistaken. First, he introduces a modified, less-violent version of the SR, which he calls the Betting-Crowd. Imagine that a casino gives even-money £100 offers to wager on a dice roll. The gambler will double his money assuming that the dice does not land double-six. Surely everybody would jump at the opportunity to take this wager. They come into the casino in batches: 1, 10, 100, 100,000 and so on. After each roll, the group must depart to give others a chance.

As ridiculous as this casino may seem, the house will always be the winner. This is because the game keeps going on until double sixes appear, at which point everybody in the casino will lose. This is a casino that takes money from 90% of all of its visitors, which is not unlike real casinos. How is this possible? Eckhardt (2013: 14) sees a direct comparison to a well-known gambling fallacy known as the martingale system; "Ninety percent of all players will lose, but I have less than a 3 percent chance of belonging to that losing majority." This may sound paradoxical, but upon closer inspection it is not.

We can also apply this thinking to Leslie's argument. For instance, we can acknowledge common assumptions regarding population growth and conclude that a large portion of people will be alive just before doomsday. This is almost a fact. However, are we justified in taking that probability and applying it to *our* situation? What is true for most people, does not have to be true for you and me. (Eckhardt, 1997; 245-6.) makes the following observation:

> If there existed a mode of statistical inference that were valid according to the extent that determinism were true … then by repeatedly testing the accuracy of this type of statistical inference, one could gauge the correctness of determinism. Since this conclusion is highly implausible, it is a safe bet that statistical inferences, including those which underlie the doomsday argument, do not hinge on the truth of determinism.

While the above statement is true, this objection assumes *repeatability* of the experiment, so it doesn't exactly strike at the core of the SR. In Leslie's original set-up, the Shooting Room is only played once, and when the dice land double-six everyone is shot.[50] If Leslie concedes that his set-up could also work in instances where the SR is repeatable, he would be cornered into taking a strange position wherein scientists could create some kind of Betting Room casino and from this alone establish whether the world is deterministic or not.

In an earlier paper, Eckhardt (1993) writes that, "as long as the validity of the Doomsday Argument is made to hinge on whether the future is open or fixed, or whether the future is fully implicit in the present, we can rest assured we are not going to settle the question of the argument's validity." We mostly agree with this sentiment. Moreover, it's not entirely clear that our world is fully indeterministic, certainly not at the quantum level. And even if it is, that doesn't mean that indeterminism is prevalent at the macro level.[51] The SR is an interesting thought-experiment and useful in showing how radical non-determinism *could* be a theoretical threat to DA

---

[50] The official reason Eckhardt (1997: 251) gives for modifying the set-up is that "in the original formulation, losing players are shot, but this added gruesomeness, if nothing else, complicates the question of how one should bet." We are not entirely convinced that the macabre nature of the SR is what motivated his alteration. Rather, the betting setting up allows for repeatability, which fundamentally changes the nature of the thought-experiment and makes it less analogous to DA.

[51] For example, uncertainty and chaos are integral aspects of our day-to-day living. This doesn't contradict a view which holds determinism at the micro level.

calculations. Nonetheless, the SR fails to deliver a fatal blow to the DA.

2.5 NORTHCOTT'S ASTEROIDS

A more recent thought-experiment forms the basis of an interesting paper by Robert Northcott (2016) and presents a unique dilemma for the DA:

> Imagine that a large asteroid, previously unobserved, is discovered to be heading dangerously towards Earth, arriving in just a few days' time. In particular, astronomers calculate that it has a 0.5 probability of colliding with us, the uncertainty being due to measurement imprecision regarding its exact path. This probability estimate is based on well-established science and is not disputed by any expert in the field. Suppose, further, that any collision would be certain to destroy all human life, and moreover that there is no feasible way to avoid it. (Northcott, 2016: 4)

We are then asked to consider how the DA should respond to such a scenario. This thought-experiment is interesting because unlike

Leslie's urns or Eckhardt's Shooting Room, we are not asked to imagine some analogous imaginary device. Instead, we are told the world is exactly as is, and are then presented with a hypothetical scenario that is not outside the realm of realistic possibility.[52] It forces the doomsayer to tackle how the DA should react to a 'sharp' empirical prediction, meaning there is very little variance in the probability density function. This is a departure from how we normally think about DoomSoon and DoomLater. Usually, when empirical estimates are discussed in DA literature, we assume that the variance of DoomSoon and DoomLater are extraordinarily large (because we are consolidating hundreds if not thousands of individual risk factors, each with high variance). Here we are essentially 100% certain regarding the 50% prediction.

Let's take some toy numbers to see why the asteroid is problematic. Assume our current birth rank of 100 billion. In the DoomSoon scenario, we reach only 200 billion humans. In the DoomLater scenario, we get to 10 trillion humans. Using Bayes formula, the ratio of *P(DoomSoon, 100)* to *P(DoomLater, 100)* would be:

---

[52] In a recent paper titled *Humanity Extinction by Asteroid Impact*, Jean-Marc Salotti (2022) calculates the probability of a humanity destroying asteroid at "between 0.03 and 0.3 for the next billion years, if there is no colonization of other planets." The study also places the probability of a massive comet impact within the next century at 2.2×10^{-12}.

$$\frac{P(DoomSoon, 100)}{P(DoomLater, 100)}$$

$$= \frac{P(DoomSoon) \times P(100, DoomSoon)}{P(100)}$$

$$\times \frac{P(100)}{P(DoomLater) \times P(100, DoomLater)}$$

The empirical prior *P(100)* conveniently cancels out, giving us:

$$= \frac{P(DoomSoon) \times P(60, DoomSoon)}{P(DoomLater) \times P(60, DoomLater)}$$

$$= \frac{P(DoomSoon) \times \frac{1}{200b}}{P(DoomLater) \times \frac{1}{10tr}}$$

$$= 50 \cdot \frac{P(DoomSoon)}{P(DoomLater)}$$

Although we don't know our priors *P(DoomSoon)* or *P(DoomLater)*, we know that conditionalizing on our birth rank, DoomSoon is now 50 times more likely (if we are considering only those two hypotheses).

Now, imagine we make the Asteroid discovery. Therefore *P(DoomNOW)* = 0.5. Doomsday is imminent, and hinges effectively on the outcome of a coin flip.[53] However, Northcott (2016: 4) points

---

[53] Technically, it would be a little bit more than 50% as there is a miniscule (but non-zero) chance that something else (e.g., a nuclear war, or another asteroid) *might* annihilate us between now and when the asteroid hits – but that probability amount is negligible.

out that according to DA reasoning, *P(DoomNOW)* is significantly *less* than 0.5. If we modify the empirically derived estimate of 0.5 by taking into account the DA figure, we get a new, adjusted figure and *P(DoomNOW)* < 0.5. This is obviously in contradiction to the astronomers estimate, and therefore incorrect. Northcott (2016: 5) writes that "the proposed DA adjustment is accordingly much more clearly implausible. We can say that it is fully *trumped* by the asteroid observation."

He provides several follow-up adjustments to the basic set-up of the experiment which aim to further elucidate the problem of trumping. For example, Northcott (2016: 5) then asks what would happen if the scientists realised that they made a mistake and actually the estimate of collision was 0.75 instead of 0.5. Do we need to gradually adjust our priors with Bayesian updating? Surely any modification would be again trumped by the newly obtained empirical estimate. What if there is another asteroid on a collision course with Mars, also occurring with 0.75 probability? We would have two scenarios where empirical evidence tells us we have identical probabilities, yet the DA suggests that because we are observers on earth, we should adjust our credence. How can the doomsayer rationalize his way out of that scenario?

2.6 CONCEPTUAL ISSUE IN NORTHCOTT'S ASTEROIDS: TRUMPING

The core insight gained from Northcott's paper is that of trumping: there may be occasions where we need to throw out DA reasoning because the newly obtained empirical evidence is so overwhelmingly strong that it renders all priors as irrelevant. Although his thought-argument indeed poses a *dilemma* for the doomsayer, [54] Northcott does provide us with an out (or three, actually), which can still salvage the DA, albeit under somewhat restrictive conditions. He points out that in the Asteroid case, we must either adjust our DA prior probabilities, or we must abandon our previous credence in favour of the new empirical DoomNOW probability, which is exactly 0.5. He frames his argument as embracing either of the two 'horns' of the dilemma. Neither option, Northcott concludes, is ideal because it requires one to either believe in the unbelievable or concede that the DA is impotent.

The third, and most interesting solution, Northcott (2016: 12) leaves for the end of his paper:

> Suppose we imagine that DA sets our prior probability for some particular date of Doom, and that now we interpret 'prior' to mean before consideration of any empirical evidence at all. It is true that, numerically speaking, incorporating

---

[54] We think it's important to note this careful choice of phrasing. Here, and also throughout other DA thought-experiment papers. It's not uncommon to come across cautious terms such as 'inconclusive', 'problematic' and 'unresolved' when examining anti-DA papers.

> DA right at the start of our sequence of calculations would yield the same results as incorporating it at any other stage. Nevertheless, philosophically speaking the difference is significant, this numerical equivalence notwithstanding. In particular, the a priori adjustment of empirical estimates falls foul of the trumping problem, but the a priori setting of priors does not. Thus, this (and only this) method of incorporating DA mathematics offers an escape route from our dilemma.

Effectively, he advocates for a reversal of the way in which we normally apply the sequence of DA reasoning. Usually, when we apply Bayes, the empirical evidence is used to adjust our prior credence's. Northcott simply asks why not let these priors already be adjusted via the a-priori considerations? There is no apparent contradiction here, as long as it is DA reasoning that motivates our priors before we take any empirical data into account. This is the only possible way we can take on the first horn of his dilemma and not encounter the trumping issue.

Even then, however, we are not fully in the clear. Northcott (2016) spends a good portion of his paper discussing the importance of (the magnitude of) the *variance* of our empirical DA priors, which will undoubtedly be enormous. This is because the estimates are drawn from a plethora of sources. Essentially, any evidence that is relevant to the survival of humanity should be accounted for (nuclear war, asteroid, alien invasion, human-eradicating AI, and hundreds if not thousands of other factors). Northcott (2016: 15) correctly reminds us in the footnote that "the literature seems to implicitly accept that, by contrast, the uncertainty of DA's probability estimates is non-negligible, which is what opens the door to empirical swamping." So only when the estimates of collision are *not* sharp (meaning there is a lot of room for error), does the DA survive. As we think this is true for now, and in the majority of the cases, we don't find this problematic. Of course, *if*, one day, we did observe something that puts an end to humanity with 100% certainty, we have to agree that would trump any and all DA considerations.

We think that there is actually another, fourth way, to think about the Asteroid, and it is not something that we encountered in Northcott's paper. What if we consider the Asteroid as a case of information asymmetry, similar to the Bob and Sally story from before? As in, the empirical predictions about the asteroid were always *out there,* we were just not aware of them. Had we had epistemic access to the information, we would have correctly incorporated that into our DoomSoon and DoomLater priors.

Consider the following simple analogy. Suzie and Arthur are betting on the outcome of a coin flip. Arthur knows the coin is weighted with a 0.9 chance of landing heads. Suzie does not know this. She uses a basic application of the indifference principle, and her credence is that there should be a 0.5 chance of landing heads. Is Suzie wrong to assume that, given she does not know that the game is rigged? No, we do not think so. She is making the most of the information at her disposal. If Arthur were to inform her that the coin is weighted, that information would indeed trump her use of the PI.

We can think of ourselves as Suzie in the framework of the DA. Perhaps there is some information out there that tells us the 'real' odds, but until we have that, we must use our best empirical estimates and assume we are randomly located amongst all observers. Imagine now that a third player, Hugo, enters and informs them both that actually, Arthur is mistaken, and the coin is in fact 0.95 weighted towards heads. This would again trump their credence, just like an updated asteroid prediction would ours. It is not controversial to think that there exists empirical evidence that would trump DA predictions. The challenge is finding what that information is, and if it is more accurate than our DA predictions. We will return to this in section 3.3.

*****

Unfortunately, the scope of this dissertation does not allow us to explore the extensive range of other DA thought-experiments and the conceptual issues that they raise. Instead, we have tried our best to distil that information in Appendix A. Nonetheless, the three thought-experiments that we have presented here all follow a general pattern that can be found in a significant majority of the thought-experiments that appear across the rich DA literature:

1. A thought-experiment is presented.
2. The thought-experiment aims to attack some aspect of the DA.
3. An 'out' is given, rendering the thought-experiment inconclusive.[55] Or, an explanation is provided as to why the thought-experiment is not analogous.
4. We learn something new about the nature of the DA, and the thought-experiment provides us with another piece to understanding the full DA puzzle.

In our analysis of the 40+ thought-experiments scattered across journal articles we are yet to encounter one that delivers a truly *knock-out punch* to the DA. This is what makes the argument so attractive. As Poundstone (2019a: 201) notes, "the Doomsday argument does

---

[55] Often this solution is provided by the same author, and sometimes within the same paper.

not fail for any *trivial* reason... It has commanded extraordinary debate because, well, that's what philosophers do...controversies get published, while consensus trivialities perish."

We concede that these thought-experiments are theoretically useful tools that raise important conceptual issues and cause headaches and dilemmas for the doomsayer. As a result, there are still many unanswered questions left, especially with regards to reference class, trumping, and determinism. Nonetheless, we believe that our best chance at refuting the DA will need to involve going beyond thought-experiments and actually tackling the DA with empirical evidence.

III. EMPIRICAL TESTING OF THE DA

3.0 SOBER'S EMPIRICAL CRITICISM OF THE DA

It is our hope that the preceding section has cast a strong doubt on the notion that thought-experiments alone can help us refute the DA. If thought-experiments can only get us so far, where else can we turn to? Elliot Sober (2003: 414; emphasis mine) suggests that "in the absence of data, we are told to follow Gott's [method]. I would expect most to say something different – in the absence of data, you should *go out and get some.*"

This, we believe, is the crucial next step to disconfirming the DA. In this section, we will show why Sober's critique works for Gott's version of the DA, however it is inconclusive with regards to Leslie's DA.

3.1 SOBER'S OBJECTION TO GOTT

Sober (2003: 415) begins his critique of Gott by restating Gott's crucial assumption:

> [P1] My present temporal position can be considered as if it were from a random sampling from S's entire existence.

He then states that this premise entails all propositions of the following form:

[MID] $P(I\ am\ in\ the\ middle\ p\ of\ S's\ existence) = p$

The probability that an observer is in the middle p of S's existence is equal to that p. This is essentially a reformulation of how we introduced Gott's method, where p was the confidence level:

t_{begin} .... p .... t_{end}

With p = 0.95,

$P(I\ am\ in\ the\ middle\ 0.95\ of\ S's\ existence) = 0.95$

Imagine now that we are in the year 2022 and the event S has lasted for 100 years. With p=0.95, the [MID] will comprise of the two extreme cases in the figures below. Note that in both cases, our current year 2022 is included within the 95% gap that is situated in the middle of S.

[CASE A]

YEAR	1922	2022		5922
percentage of	↓	↓		↓
S's timeline	[--2.5%--][--------------------------95%--------------------------][--2.5%--]			

[CASE B]

YEAR	1922		2022	2024.6
percentage of	↓		↓	↓
S's timeline	[--2.5%--][--------------------------95%--------------------------][--2.5%--]			

CASE A puts 2022 directly at the start of this middle interval and CASE B puts 2022 at the very end. This shows us that if S is 100 years old and the current year falls inside the central 95% of S, then S will continue between 3900 [= 39 x 100] and 2.6 [= 100 / 39] more years. Sober (2003: 418) then correctly recognizes that Gott's sampling assumption has further significant consequences.

Specifically, premise [P1] would also mean that:

[FIRST] $P(I\ am\ in\ the\ first\ p\ of\ S's\ existence) = p$

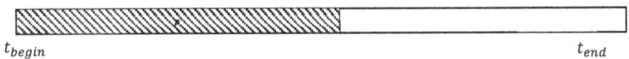

t_{begin} $\qquad\qquad\qquad\qquad\qquad\qquad\qquad\qquad\qquad$ t_{end}

And

[LAST] $P(I \text{ am in the last } p \text{ of } S\text{'s existence}) = p$

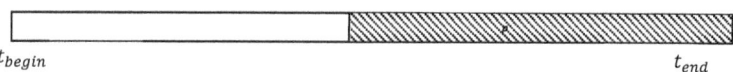

t_{begin} t_{end}

Essentially, we can think of this as sliding that confidence interval to the two extremes. Sober (2003: 418) then goes on to point out that when we *conjoin* these implications with [P1], we get some interesting predictions.

For instance, by setting p = 0.9 :

$$100 \cdot \frac{\left(\frac{1-0.9}{2}\right)}{0.9 + \left(\frac{1-0.9}{2}\right)} < t_{future} < \frac{\left[0.9 + \left(\frac{1-0.9}{2}\right)\right]}{\left(\frac{1-0.9}{2}\right)} \cdot 100$$

$$5.263\ldots < t_{future} < 1900$$

Thus, for MID-90, we can assert that there is a 0.9 chance that S will last between 5.26 and 1900 years. We can keep doing this for different values of p and collect the results in the following table:

TABLE 1
Probability distribution of S

Probability	Additional Years	
x 100	(For an S that began 100 years ago)	
0.99	0.5	19,900
0.95	2.6	3900
0.9	5.26	1900
0.5	33.3	300
0.2	66.6	150
0.02	96.1	104.1

Sober (2003: 419) then notes that when Gott uses this method on two of his favoured examples (the Berlin Wall and the USSR), he obtains a singular interval. As this interval is rather wide due to the 95% confidence level, those two events expired well within the range.

> This is like testing the hypothesis that a coin is fair by seeing if it produces between 1% and 99% heads in a hundred tosses. The hypothesis *does* make that prediction, and the fact that the coin behaves as predicted is of *some* significance. But the prediction is very weak and unspecific, and the confirmation this confers on the hypothesis is correspondingly meagre. A better test of the hypothesis would look

at the hypothesis' more specific predictions. (Sober, 2003: 419)

The critical question is therefore the following. Do we have good empirical reasons for thinking that the Berlin Wall, the USSR, Broadway plays, romantic relationships, and most importantly, the survival of our species, all follow the specific sampling distribution assumed by Gott? Sober's (and our) retort is a firm no, as we are rather confident that the empirical evidence would not agree with these frequency tables. For instance, in 1969 (the time Gott made his prediction) the Berlin Wall had already stood for 8 years. We could construct a similar distribution table:

TABLE 2
Probability distribution of fall of Berlin Wall

Probability x 100	**Additional Years That The Berlin Wal Will Last**	
0.99	0.04	1592
0.95	**0.21**	**312**
0.9	**0.42**	**152**
0.5	2.66	24
0.4	3.43	18.6
0.2	5.3	12
0.02	7.7	8.3

Although Sober never quite explicitly states it, he in effect extracts the following interpretation from Gott's DA:

> *For any S that has a confidence level set at $p = x$, all instances of S that fall outside this interval will do so with a probability of $1 - p$. Where S is an event in which we are randomly located observers, and $0<p<1$*

For example, would we expect one out of every twenty walls *just like* the Berlin Wall to last somewhere longer than 152 years?[56] Or one out of every forty to stand beyond 312 years? This is the reference class issue (Type-II), once again. Sober (2003: 419) maintains that no, "I doubt that these barriers exhibit the distribution of longevities that Gott's sampling assumption predicts."

We suggest taking this analysis even further and seeing what the implications would be for another one of Gott's predictions: current romantic partnerships. Assume that a randomly observed couple has been together for exactly 1 month.

---

[56] Sober (2003: 419) correctly points out that we should not really be comparing the Berlin Wall to regular 'walls' in general, but instead to 'barriers' that separated areas and prohibited travel.

TABLE 3
Probability distribution of romantic relationships

Probability x 100	Additional Months Relationship Will Last	
0.99	0.005 months (~ 3.5 hours)	199 months (~ 16.5 years)
0.95	0.026 months (~ 18 hours)	39 months (3 years + 3 months)
0.9	0.053 months (~ 39 hours)	19 months (~ 1.5 years)
0.5	0.333 months (~ 10 days)	3 months
0.4	0.43 months (~ 13 days)	2.33 months (10 weeks)
0.2	0.666 months (~ 20 days)	1.5 months (6.5 weeks)
0.02	0.96 months (~ 29 days)	1.04 months (32 days)

Is it really feasible to expect that only one out of every forty relationships that are today celebrating their one-month anniversary will go on to last just over three years? Is it reasonable to think that only one out of ten of these relationships will last longer than nineteen months? We do not think that the empirical evidence would support this claim. Moreover, cultural differences, legal barriers regarding marriage, and temporal considerations surely *must* play a role in predicting how long a romantic relationship will last.

It is hardly debatable that a one-month long partnership at the start of the 1700s would have a wildly different expected length than a one-month long romantic relationship today. Gott's method does not allow for this additional information to be incorporated.[57] It clearly predicts that *all* one-month long couples throughout all of history would have exactly the same odds of living happily ever after. Thus, once again, the trumping issue and the reference class issue rear their ugly heads.

If we are now seriously doubtful that Berlin Wall-like barriers, dictatorial regimes, and romantic relationships all fail to empirically satisfy Gott's sampling assumption, why should we expect the lifetime of our species to be any different? Alas, we do not have reliable actuarial data on human-like species that have come and gone, however an empirical induction against Gott's method is still possible as he makes some interesting predictions about the lifetimes of other species.

Gott's argument is indirectly making the claim that species that have been around for a long time will survive longer than species that have existed for a shorter time.[58] In the context of the survival of different animal species, this is simply not true. Moreover, it goes against established theories of evolutionary biology (see Dawkins and Krebs, 1979). For example, Van Valen's (1973) 'Red Queen Hypothesis' argues that a species chances of extinction are *independent* of how long that species has been around. Instead, Van

---

[57] Unless, of course, you differentiate 'modern relationships' and 'early 1700s relationships' with qualifiers. This is the reference class Type-II issue once more. d

[58] This is clear when we compare the distribution tables for an S lasting 1 month, 8 years, and 100 years. It is also evident from a quick inspection of Gott's equation.

75

Valen argues that species need to continually change and adapt in order to survive in a constantly changing world. The better a species can adapt, the better the chances of survival. Any *a-priori* speculation about a species total lifetime is futile because we have an abundance of empirical data that disconfirms Gott's prediction. Who should we listen to? The philosopher-doomsayer armed with some a-priori estimates based on the Copernican Principle? Or decades of empirical evidence carefully studied and dissected by the world's leading biologists?[59]

It is also rather worrying that Gott's delta-t argument leads to another paradoxical result if applied to the survival of any *individual* member of any species. For example, a straightforward application of Gott's framework to predict how long a 30-year-old adult will live would lead to a result where they should live more additional years than a 1-week-old baby.[60] Of course, this is absurd. The older an organism is, the sooner they will probably die. This is another instance of the trumping issue. For any given individual member of a species, prior duration is *negatively* correlated with the number of years it has left to live.

A secondary concern is that Gott could just keep adjusting his confidence level so that a species extinction date does fall in between

---

[59] It should be noted that Professor Sober's expertise lie primarily in the philosophy of biology.
[60] This has been coined the 'Baby-Paradox'. See Delahaye (1996) and Korb and Oliver (1999; 405).

the predicted period. Note that setting a p value of *almost* 1 in Gott's equation would mean that the species will go extinct within a period of (almost) now and up to (almost) infinity:

$$t_{past} \cdot \frac{\left(\frac{1-p}{2}\right)}{p+\left(\frac{1-p}{2}\right)} < t_{future} < \frac{\left[p+\left(\frac{1-p}{2}\right)\right]}{\left(\frac{1-p}{2}\right)} \cdot t_{past}$$

This kind of prediction is too weak to be considered of importance.[61] Moreover, recall how Gott's sampling assumption implies that species that have been around longer will outlive (in terms of species extinction) 'younger' species. There are some species alive today that have been around for hundreds of millions of years: [62]

TABLE 4:
Different Species Ages of Prehistoric Animals

Species	Species Age Today (years)	Gott's Prediction (years) [p=0.95]
Sponge	760,000,000	$19{,}487{,}179 < t_{future} < 29{,}640{,}000{,}000$
Jellyfish	505,000,000	$12{,}948{,}717 < t_{future} < 19{,}695{,}000{,}000$
Horseshoe Crab	445,000,000	$11{,}410{,}256 < t_{future} < 17{,}355{,}000{,}000$
Sturgeon	200,000,000	$5{,}128{,}205 < t_{future} < 7{,}800{,}000{,}000$
Goblin Shark	118,000,000	$3{,}025{,}641 < t_{future} < 4{,}602{,}000{,}000$
Homo Sapiens	**200,000**	$5{,}128 < t_{future} < 7{,}800{,}000$

---

[61] This could be construed as a form of p-hacking (incrementally altering statistical analyses until nonsignificant results appear to be significant).
[62] Species age figures taken from Ciampanelli (2015).

Are we really to believe that goblin sharks will outlive the human species by a factor of 600? Or that jellyfish will be around billions of years after us? Surely many scenarios that would wipe out the human race (e.g., nuclear annihilation, asteroid, heat death of the universe) would also wipe out some (but perhaps not all) of the above-mentioned species? Gott's method is clear that the older species that have existed since the time of dinosaurs have a much better chance of out-surviving younger species such as ourselves. There is nothing in evolutionary biology that would explain such a prediction. In fact, most of the current theories are in stark contradiction to this claim.[63]

We should also ask ourselves if making comparisons to other species' survival is actually a useful analogy to make at all when discussing the DA. This is the Type-II reference class issue again. Indeed, we are talking about extinctions of animals in both cases, but we are in fact the only species to be exposed to a plethora of (man-made) existential risks. We are the only species to have developed nuclear weapons, biological warfare, nanotechnology, and AI that could potentially kill us. Thus, it seems a bit odd to consider how other non-human species have gone extinct when discussing our own fragility and longevity. Especially since it was

---

[63] It is also an inescapable truth that the Sun will explode in 5 to 8 billion years.

us, humans, who were responsible for most, if not all, of their extinction.

Gott's reliance on the CP to justify his sampling assumption is only plausible if we lack any other relevant empirical information. This is not the case for the vast majority of things we are interested in looking at. Even taking Gott's favourite duo of famous walls and romantic relationships, we fail to satisfy the assumption. If Gott's method makes empirically *weak* predictions for the events that he claims are a good fit, then we have very little reason to believe that using it to forecast the longevity of our species would result in a significantly better prediction.

3.2 SOBER'S OBJECTION TO LESLIE

Finding Sober's critique of Gott more than compelling, we can now turn our attention to how he tackles Leslie's DA. First, Sober (2003: 422) recognises that there is yet another (fourth) crucial aspect in which the two versions of the DA differ, noting that "Leslie's is not an argument about what will *probably* happen… It is an argument about likelihood". He therefore proceeds to reframe Leslie's DA as following the likelihood principle (see Hacking, 1964):

> Likelihood Principle [LP]: O is evidence for H_1 over H_2 iff $P(O|H_1) > P(O|H_2)$

In any Bayesian framework, the likelihoods of two rival hypothesis H_1 and H_2 are related in the following manner:[64]

$$\frac{P(H_1|O)}{P(H_2|O)} = \frac{P(O|H_1)P(H_1)}{P(O|H_2)P(H_2)}$$

It is clear from the above equation that the ratio of *posterior* probability will be more than the ratio of the priors if H_1 has a greater likelihood than H_2. As before when we covered the example in Northcott (2016), we can rephrase Leslie's DA as a ratio to obtain:

$$\frac{P(DoomSoon|Br)}{P(DoomLater|Br)} = \frac{P(Br|DoomSoon)\,P(DoomSoon)}{P(Br|DoomLater)\,P(DoomLater)}$$

Sober's rationale for treating Leslie's DA as an argument about likelihoods is rooted in Leslie's description of the two-urns experiment and also his repeated insistence that his DA, "strictly speaking... is only for a *shift* in any estimate of the risk of our race's imminent extinction" (Leslie, 1996: 70).

Recall that Leslie also makes the following assumption:

---

[64] This is obtained by a straightforward division of the two Bayes formula – the bottom terms cancel out.

[A1] My birth can be treated as if it was the result of random sampling from set of all humans who will ever exist.

This is evidenced by Leslie's reliance on the two-urns experiment and his insistence that the scenario is analogous to ours. From this assumption (and the above likelihood principle) Sober (2003: 423) obtains the following inequality [I]:

$$P(Br|DoomSoon) > P(Br|DoomLater)$$

Essentially this says that the probability that I exist now (conditionalized on DoomSoon) is *larger* than the probability that I exist now (conditionalized on DoomLater). To help conceptualize this, consider the following diagram, based on Sober's (2003: 423) sketch:

DIAGRAM 3

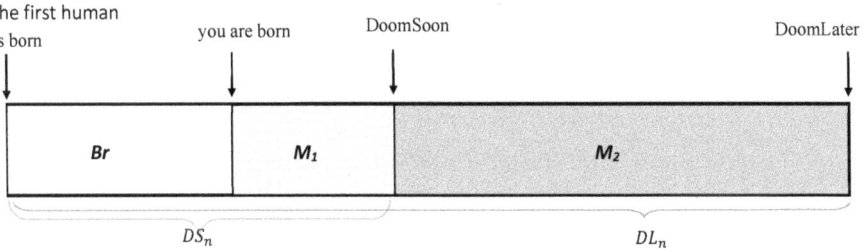

Recall that previously we denoted the probability of having your specific birth rank in the DoomSoon scenario as

$$P(Br, DoomSoon) = 1/DS_n$$

In the above graph, $DS_n = Br + M_1$

Similarly, the probability of having your specific birth rank if the DoomLater scenario is true was given as

$$P(Br, DoomLater) = 1/DL_n$$

Which, shown in the graph above is $DS_n = Br + M_1 + M_2$

Thus, the inequality [I] can be expressed as

$$P(Br|DoomSoon) > P(Br|DoomLater)$$

$$\frac{1}{DS_n} > \frac{1}{DL_n}$$

$$1/[Br + M_1] > 1/[Br + M_1 + M_2]$$

Intuitively, this makes sense as long as assumption [A1] is correct. If you randomly sampled from the set of all humans who ever existed, the reciprocal of that will be larger when the denominator is smaller $[DS_n < DS_L]$. It is also important to once again note that historical population growth rates do not factor into Leslie's DA, in neither the mathematical exposition nor the diagram above.

Sober (2003:423) then says we can further extrapolate the following from the above inequality:

P[I exist|doomsday occurs in x years)
$> P[I\ exist|doomsday\ occurs\ in\ y\ years)$

For all $x < y$

Moreover, Sober (2003: 424) points out that Leslie's sampling assumption [A1] clearly conflicts with Gott's assumption [P1]:[65]

> [A1] My birth can be treated as if it was the result of random sampling from set of all humans who will ever exist.
>
> [P1] My present temporal position can be considered as if it were from a random sampling from *S*'s entire existence.

Therefore, only one of them can be correct. Or as Sober attempts to argue, neither are correct. Note that Leslie wants us to reason as if we

---

[65] The one exception to this would be if people were born in a perfectly uniform distribution over the entire lifespan of human existence. Of course, we already know this is not the case.

were drawn from a system that uses a uniform distribution (like balls coming out one by one from the urn). Sober (2003: 424) is right to question exactly *what kind* of distribution Leslie has in mind, noting that, "we could draw from time intervals of equal duration, assuming that they are equiprobable, or we could draw from the list of birth dates occupied by human beings past, present, and future, on the assumption that these are equiprobable".

Moreover, as already discussed in 2.2, our future population growth remains uncertain. Consider the following population models, all of which are plausible as long as the left section is in line with the exponential population growth of the last millennia:

DIAGRAM 4: POTENTIAL FUTURE POPULATION DISTRIBUTIONS

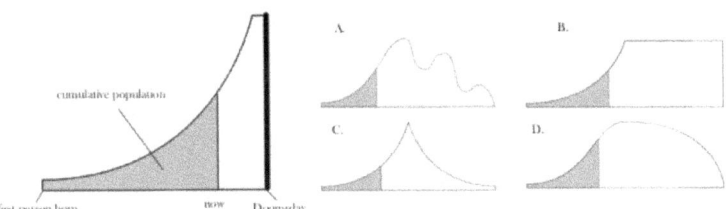

We can certainly see how this poses a problem for Gott's DA, as we would be required to have actuarial knowledge of future demographic data (alas, if we had that, we would not need the DA). However, we do not think this distribution poses a direct problem for Leslie, as Sober seems to suggest. All Leslie really says is it is extremely improbable that we are early in our existence. The uncertainty regarding the distribution only really affects the magnitude of the shift (which is obviously sensitive to the population growth in the future). There is, therefore, a kind of trade-off. The future lifespan of humanity could be lengthy but only if it is not populous. Or the future could be heavily populated but only if doom is around the corner. Leslie's point is that we are very unlikely to reach an absurdly large cumulative population.[66]

In his attempt to disprove Leslie's DA, Sober (2003:424) focuses his attention on the crucial assumption [A1], insisting that our "present temporal location is *not* the result of random sampling from the

---

[66] Whether or not is this is a vacuous claim will be assessed in Part IV.

temporal locations of all human beings". To illustrate why, Sober attempts to rephrase assumption [A1] in terms of expected value [EV]. He relegates his definition of EV to a footnote, stating that his "talk of 'expected value' of a quantity refers to the mathematical expectation – the average value that would arise under repeated trials".[67] We are provided the following schematic:

DIAGRAM 5: EXPECTED VALUES

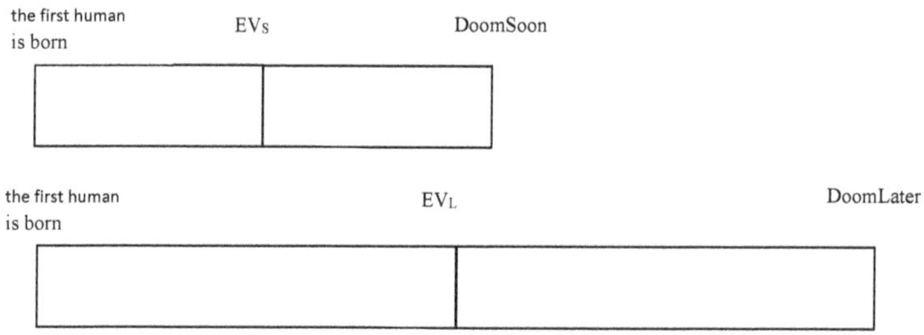

the first human is born EV_S DoomSoon

the first human is born EV_L DoomLater

Sober (2003: 425) then argues that in the DoomSoon case, your 'expected value' would be exactly "in the middle of the timeline", shown as EV_S. In the DoomLater scenario, it would be EV_L. He provides no explanation as to why this would be the case, stating only that "since my temporal location is to be assigned randomly, it could be anywhere – however, its expected value is in the S's duration".

---

[67] It is our understanding that what Sober has in mind is the EV of a 'hypothetical' random draw from a sample of potential birth rank values. Therefore, *in the hypothetical space*, the draw doesn't only happen once.

This, as we understand it, would only be the case if Leslie assumes a distribution where the population growth rate is constant, and you have an equal likelihood of being born at any point along the timeline. But Leslie makes no such assertion. Again, it is not clear how this EV can be worked out (as we do not exactly know how many humans would exist in the possible scenarios).

Sober (2003:425) then claims that if we had actuarial data for a list of species that we belonged to (it's hard to comprehend how someone can belong to more than one species) that we should expect, on average the species we belonged to sooner, to have an earlier doomsday. He claims that the above analysis "predicts that the doomsday of these different species should be correlated with the times at which I happen to belong in them" (2003: 426). We fail to see why he makes this claim, and also why he would consider 'belonging' to a species as something analogous to being born. This is even more apparent in the following troubling paragraph:

> Although I belong to just one species, I have joined different organizations at different times in my life. In childhood, I joined the American Numismatic Association. As a teenager, I was in an orchestra. When I first came to Wisconsin, I joined a sailing club. Now, in the wisdom of my maturity, I am a card-carrying member of the American

> Philosophical Association. Suppose these organizations all began at the same time and that each is still around now. When will each exit from the scene? Leslie's likelihood inequality, applied to these organizations, predicts a correlation. These organizations should go extinct sooner, the earlier in my life I joined them. If Leslie is right, this is bad news for the ANA, but good news for the APA. (Sober, 2003: 426)

It is unclear how being born into the totality of all humans is in any way akin to joining and becoming a member of an organization. While we appreciate that Sober is primarily concerned with the underlying sampling process, it seems he may be equivocating when claiming what Leslie's DA actually implies. He summarizes his objection in the following crucial paragraph:

> Surely the general pattern is that if I bear relation R to x before I bear R to y, where x and y both exist now and are equally old, there is no general tendency for x to go extinct before y... Leslie's as-if sampling assumption predicts that there should be a correlation between when I exist and when the human race will go extinct. This correlation is to be

> expected if my birth sets a time bomb ticking that eventually blasts the human race to oblivion, or if the birth and the extinction are joint effects of a common cause. (Sober, 2003: 427)

It is here where we find that Sober's critique misses its mark. An argument can certainly be made that by being born, you ultimately *contribute* to the demise of humanity. There is a fundamental difference between joining different organisations or joining the list of all humans who will ever exist. To understand why, consider our following simple thought-experiment which we call *The Densely Packed Island*:

Imagine there is only one populated island on earth, and it has ten humans living on it. There are only a set number of existential risks that could kill these ten people. Let's assume that there are only four things that wipe them out: volcanic eruption, starvation, mass drowning, and disease. As more people are born, eventually there will be more risk factors. Some of the aforementioned risks now become more likely (e.g., starvation if resources are limited, or disease if the island becomes too densely packed). Perhaps the islanders develop weapons of mass destruction, or eventually AI which becomes dangerous. This has all become possible because of contributions made by new members of the population. While we understand the point Sober is ultimately trying to argue, it does not make sense to draw a comparison to joining different organisations because the situations are not analogous enough.

In that sense, each person joining this 'organization of humans' *does* set off a sort of metaphorical time bomb. This densely packed island with more and more existential risks can be seen as a microcosm for our situation here on earth. Each person who is born into the world doesn't set off a literal time-bomb themselves, yet they may contribute. Each additional human increases the risk of extinction. [68] Your birth hasn't 'set a time bomb ticking' but you may have seriously contributed to something that would.[69] For example, each individual born has a carbon footprint which contributes to global warming.

The obvious limitation to this reply is that it is rather *ad-hoc*. There is nothing in the fundamentals of Leslie's DA or of Bayesian reasoning that mentions or incorporates growing existential risks. The DA makes no appeal to empirical justifications as to why an increased population would lead to increased risk. The DA's strength is that it is supposed to work because of probability and probability alone.

---

[68] When we raised this objection with Sober (personal communication), his response was that "it depends on what these additional humans are like. Some may turn out to be evil doers who promote climate crisis and war, and *that* would increase the probability of extinction. However, others maybe do-gooders, who will have the opposite effect, possibly by frustrating what the evil doers set out to do." Implicitly, it seems as if Sober also accepts the trumping objection.

Still, we are not convinced that Sober's general objection to Leslie holds. Indeed, it *could* be that there is no correlation between when one exists and when the human race will die out. However, Sober hasn't provided solid evidence for that being the case (and, in fairness, Leslie hasn't provided evidence to the contrary). The onus falls on Sober to provide proof why it is not so, as he is the one making the objection. For these reasons, we conclude that Sober's critique of Leslie's DA is, at best, inconclusive.

3.3 TESTING THE DA ITSELF VIA SIMULATION

Perhaps we can actually do better than testing the underlying sampling assumptions and go directly to testing the DA itself. Is the DA testable and falsifiable? For Gott, the DA is falsifiable if it can be shown that more than 2.7 trillion humans will be born. A sentiment that is commonly repeated in the DA literature is that we will not have an answer to that question until doomsday actually occurs. The obvious paradox here is that there will be no observers left to make that observation. Consider the following quotations:

> "… It's more accurate to say that the doomsday argument is unverifiable… we can't be certain the prediction is right until, uh, everyone is dead." (Poundstone, 2019a: 66)

> "We do not even know if there should exist some extremely dangerous decay of say the proton

> which caused the eradication of the earth, because if it happens we would no longer be there to observe it and if it does not happen there is nothing to observe." (Nielsen, 1989: 452)

> "… Of course, I am a member of just one species, and there is just one of me, so this test cannot be conducted." (Sober, 2003: 426)

We think that the above statements could be contested. Consider the very real possibility that we will one day be able to model our entire universe and run complex ancestor simulations.[70] If we are then able to speed up these simulations, we would, in theory, be able to see when and how often doomsday occurs.[71]

For example, assume we are able to run a million simulations starting at present-day all with the exact same starting conditions. If we forward-test these scenarios, we will get a range of results. In the graph below we have sketched up some possible timelines for various simulations ($SIM_1 - SIM_{1million}$). You can see doomsday occurring at different times:

---

[70] Ancestor simulations are just computer-generated replications of our ancestor's history. Whether this will be technologically and philosophically possible is examined in Bostrom (2003). For a critical assessment of the Simulation Argument, see Zouev (2019).

[71] Fast-forwarding time is already a key feature of simulation games and software like *The Sims* and *Sim City*. The same idea would apply here but on a much larger scale.

DIAGRAM 6: SIMULATING DOOMSDAY

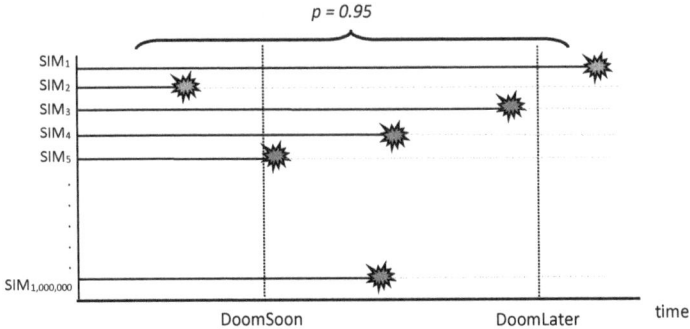

Testing Gott's prediction with this method would be rather straightforward. We just need to see how often the simulated human civilizations reach doomsday and if this figure falls outside Gott's confidence level (set at p=0.95). For example, if we have 100 simulated universes and *only 3 of them* last less than 12 or more than 18,000 years then we can say that Gott's 95% level is satisfied as 3% is less than 5%.

In order to test Leslie's DA by way of simulations, this approach would be less straightforward as Leslie does not actually provide a specific prediction. However, if we keep to the example of two possibilities, DoomSoon and DoomLater, we can nonetheless check to see how close we were to either prediction. If most simulations came before DoomSoon, then this is evidence that the probability shift predicted by Leslie is real and could be interpreted as confirmation of the DA. Alternatively, we could also run these simulations using various different start dates, as well as focus on other species or events.[72] This way we can back-test if the correlations predicted by Leslie and Gott are indeed correct.

This proposed solution also has the added benefit of overcoming another DA objection: you cannot draw conclusions from a singular event (Poundstone, 2019a; 69). Having millions of futures play out will allow us to get a truer probability estimates for our own doomsday. Of course, such conjecture is speculative at best (and implausible at worst) as we currently lack the computing power required and are nowhere near to developing the technology needed to run these fine-grained and detailed life simulations. However, we do already have a variety of risk models and hazard-specific simulations that show us, for example, what will happen with regards to climate change as well as models for simulating nuclear war and pandemics. If we were able to aggregate these models, we could get close to an answer without having to build an actual aggregated simulation of existential risks.

However, it appears to us that a strange paradox arises in this discussion of estimating doomsday risk. Surely there are some models where the number of humans who have ever existed is already, in some sense, *built into the estimate*. For example, climate change

---

[72] I thank my supervisor for pointing out this simple but effective suggestion.

models would factor in that we are in the year 2022 and there have been 300 million humans so far. If this is the case, is there not a worry that we would 'double-count' evidence when applying Leslie's probability shift? The only place we have seen this matter mentioned in the literature is in Poundstone's (2019a; 183) book, emphasis mine:[73]

> We therefore need to consider exactly what these "prior probabilities of early doom" are. In 2003, Martin Rees created a stir by estimating the chance of civilization surviving the twenty-first century at only 50 percent. He based this estimate on his assessment of the risks of global war and nuclear, biological, and nanotechnological terrorism (not factoring in the doomsday argument)... Should we then use our position in time to adjust Rees's estimate, bringing doomsday even closer? No. Rees's pessimism most definitely comes with a time stamp. He was speaking of uniquely

---

[73] We are not the only ones to have this concern. In a brief email exchange with prominent Russian transhumanist and DA proponent Alexei Turchin, he informed us that he, too, had this worry. Turchin has authored several books on calculating global existential risks (see Turchin, 2010, 2015) and replied that (translated), "when I was writing *The Structure of Global Catastrophic Risk,* I was also worried that risk estimates we plug as a priori into Leslie's formulation would lead us to 'double counting' regarding our position - and this is not good! The solution would be to take as a-priori the 'pure' probabilities of, say, nuclear war or asteroid etc., *without* taking into account our anthropic information that it will happen. But nobody really knows what those values are..."

twenty-first-century risks, those tied to technology and population that are without precedent in human history. This is not Rees's idiosyncratic belief. Anyone who thinks about it must agree that existential risk is not constant with time. When scholars or think tanks estimate existential risk, they are of course considering our moment in time. *It follows that I can't pick out a prior probability applicable to my epoch and then turn around and adjust that prior for my position in time. That would be double-counting evidence. Any well-informed assessment of existential risk already incorporates our position in time.* That leaves little or nothing to be learned from doomsday reasoning.

This is a concern to which we currently do not have a satisfactory answer.[74] Interestingly, however, we could avoid this worry if we makes DA calculations in the manner as advocated by Northcott (2016: 12) in the penultimate section of his paper. That is, if we do the DA calculations at the *very start* and use them to inform our priors, then this issue of double counting is potentially avoided. As aforementioned, this would be the opposite way to how Leslie

---

[74] Personal correspondences with both Nick Bostrom and John Leslie on this exact issue were not particularly fruitful. However, that was likely due to a difficulty in us conveying what exactly we mean by this issue.

advocates DA calculations as he always *first* incorporates all the empirical data and *then* lets the DA influence the priors.

There is also a secondary paradox that we may encounter in our proposed approach. One that concerns the incoherence of simulated predictions and how they can be self-defeating. For example, imagine we keep seeing that nuclear war destroys humanity in 99% of these simulated timelines, we would then take immediate action today with regards to full nuclear disarmament (one would hope, although this is certainly not a given). However, if we did that, then the simulations do not really give us a true prediction of the future – unless there are cases where the simulated worlds also abolished all nuclear weapons. Thus, the pressing question is when and why such predictions are indeed self-undermining (or self-fulfilling), and when they are not. Perhaps there is a further possibility that human awareness of the simulation's results could *itself* be programmed into the simulation. It is probably too premature to speculate on such matters, however we do feel these concerns will be relevant in the future. If, that is, our species survives long enough to be able to run such simulations.

*****

It might be of interest to now revisit our trio of conceptual issues and briefly consider how a simulated world could help us with some of our unanswered questions.

Reference Class in the Simulation. The worries about correctly choosing a suitable reference class would largely drop out if we were able to run simulations. In fact, we could use them to see for which reference classes Gott's or Leslie's DA would have matched empirical predictions. For example, if we could recreate our world in many different simulations, we would be able to see timelines of different species and get better data on predicting their longevity. This would help us solve Type-I reference class issues.

Furthermore, we could also easily overcome Type-II reference class issues. If we could simulate entire worlds then it would be a very straightforward exercise acquiring actuarial data on the lifespans of different organizations, famous walls, romantic relationships, Broadway plays, and literally everything and anything Gott was interested in. It would then be very easy to test his delta-t hypothesis. Moreover, we would have a clearer answer to Elliot Sober's question as to whether the underlying sampling process conforms to Leslie's, Gott's, or (as Sober himself speculates) neither of their assumptions.

Trumping in the Simulation. It us our opinion that trumping will still be a major concern, even if we are able to simulate entire timelines. For example, let's say we run a thousand simulations and we get an empirical estimate for what is most likely to kill us and when. The critical question remains: how much epistemic authority should we assign to these predictions?

All simulations would in effect be empirical (or at least semi-empirical) as they would be constructed from observations we make in the real world. Are these estimates really *superior* to the empirical priors we normally plug into the Leslie's DA formula? How different are they going to be?[75] Should we combine the two independent estimates? Perhaps in a way advocated by Northcott (2016: 7) by using the inverse-variance weighting? To adopt the estimates obtained from a simulation would be to implicitly concede, as Northcott warned us, that empirical evidence trumps a-priori. Thus, the problem of trumping not only remains, but the obtained simulation results may actually prove trickier to dismiss than a hypothetical asteroid.

There is also another worry. Let's assume we are able to run a thousand highly detailed world-simulations, but then, one day, those astronomers from Northcott's thought-experiment come to us bearing the same bad news.[76] Despite whatever the computer models tell us, we still have an asteroid coming at us with a sharp 0.5 chance of

---

[75] We do think it is a safe assumption to make that simulating future outcomes by actually rebuilding the entire world is going to give us better predictions than the more informal analyses we already have about the future of humanity (e.g., Rees, 2003). We concede though that this claim requires further scrutiny.

[76] There is another nuance here. How do we build a simulated model that includes something exogenous to our system like an asteroid from outer space? We could have a random variable but what would this be based on? Probably also our best empirical estimates.

killing everyone.[77] We may very well end up with a situation where we have (at least) three different probabilities for DoomSoon. First, we have the usual empirical priors (as discussed above, by Rees for example), $\text{DoomSoon}_{\text{Empirical}}$. Then we have the probability that was obtained by running thousands of simulations, $\text{DoomSoon}_{\text{Simulated}}$. Again, we would have a sharp empirical prediction from the astronomers, DoomNow, which DA reasonings tells us we must adjust. How should we respond? It is not entirely clear. What is clear, however, is that the challenges of which horn to embrace reappear and the issue of trumping lingers on.

We don't even necessarily need to simulate an actual asteroid collision to drive home the point. Consider the final variation of Northcott's (2016: 6) thought-experiment:

> Imagine a set-up in which the fate of Earth is made dependent on the outcome of a single indeterministic quantum event. Suppose in particular that this event will yield, say, either an Up or Down result, each with 0.5 probability, and that things are so rigged that Up would trigger Earth's demise while Down would spare us. Therefore, at the moment just before the event, Doom-Now has a probability of 0.5. Nevertheless, DA

---

[77] This has an element of the SR as well. A computer told us that 99% of worlds like ours will survive, but our best scientists are now saying our destiny hinges on a coin flip.

reasoning would still urge us to reduce our credence in Doom-Now, and hence in Up, to less than 0.5, even though that 0.5 value is the direct implication of well-confirmed physical law. Thus, our credence here would contradict a law of nature.

We can already more or less do this with the current simulation models we have.[78] Input a parameter where if Up, the game shuts off, and if Down, the game goes on. Using DA reasoning to estimate future longevity of the individuals in that simulation would *have to* be superseded by this simple physical fact.

Determinism in the Simulation. Would simulations help us get an answer to Leslie's question about a radically indeterminate world? We believe so, as the simulations would be able to show us that despite exactly the same starting conditions and parameters, the simulated timelines took vastly different paths.

Simulating the SR itself could be relatively straightforward and has been alluded to by Eckhardt (2013: 12) within the framework of his modified Betting Crowd. Similarly, if we run many simulations of the

---

[78] We are thinking here of something like the latest version of a highly detailed life-simulation game such as *The Sims*. However, this argument would also apply to even the most rudimentary of simulation games, such as Conway's *Game of Life*.

SR and focus on one player (assuming they 'respawn' each time they are killed)[79], we will find that they would only die one in 36 times.

Perhaps here would be a good place to comment on the type/token distinction inherent in doomsday simulation probabilities. Simulating the chances of dying in the SR (or humanity ending in entire-world simulations) would give us *type*-probabilities. In effect, we obtain probabilities over the sample space which we got from running thousands of simulations.

The problem is that we do not have the benefit of repeatability in the real-world. Something like Northcott's asteroid, for example, would be a *token* probability, in the sense that it's actual and not derived from repeated experiments. Given the choice, should we update our credence using type or token probabilities? Surely the rational answer is that we use the real-world token probability.

---

[79] Granted, this is a strange assumption as we only have one life in the real world. This aspect of *repeatability* which Eckhardt's conveniently smuggles into his variation of the SR changes the dynamics of the thought-experiment considerably.

IV. IMPLICATIONS OF THE DOOMSDAY ARGUMENT

Despite the cynical moniker, the DA does not predict that doomsday is imminent, and nor does it suggest that it would be pointless to try and change the course for humankind (especially if one is sympathetic to Northcott's trumping critique). On the contrary, the role *P(DoomSoon)* and *P(DoomLater)* play in Leslie's formulation would suggest that any efforts to change these priors could have an immense impact on our predictions. Mitigating the dangers of global warming or nuclear war would decrease the prior probability of DoomSoon, which would then translate to a reduction in the posterior probability of DoomSoon once Leslie's DA is accounted for. The DA appears to suggest that reducing existential threats to the survival of our species should be even more imperative.

Moreover, in the event of a drastic decline in world population, it would take much longer before there would have been enough additional humans born to make your birth rank look surprisingly low. This is a notion that is seldom discussed in the more pessimistic DA journal articles. Therefore, a nuclear holocaust (or any mass cataclysmic event), which *all but* wipes out the human race and leaves only a few survivors is actually something that would instantly guarantee a massively decreased future population, and, in terms of the DA, would be 'good' for us because we are therefore not exceptionally early.[80] However, having 99.99% of the population instantly perish is not significantly 'better' than total extinction in terms of the ethical implications, yet the DA does not really accommodate this idea.[81]

As we already discussed briefly in the introduction, even if the DA is correct, other interpretations could be possible. For example, perhaps we will evolve into 'posthumans' who would fall under a different reference class. In this case, the DA would merely suggest that some kind of evolution would occur before there were immeasurably more humans born. Or perhaps we will colonize the galaxies and live on infinitely – in which case the DA calculations are inconclusive.[82] A more precise explanation of what the DA demonstrates is, therefore, as "a disjunction of possibilities rather than as the simple statement: Doom will probably strike soon" (Bostrom, 2002a: 213). Even this,

---

[80] I thank my supervisor for pointing out this morbid fact.
[81] This notion has been contested. Consider Derek Parfit's (1984) thought-experiment:
"I believe that if we destroy mankind, as we now can, this outcome will be much worse than most people think. Compare three outcomes:
(1) Peace.
(2) A nuclear war that kills 99% of the world's existing population.
(3) A nuclear war that kills 100%.
(2) would be worse than (1), and (3) would be worse than (2). Which is the greater of these two differences? Most people believe that the greater difference is between (1) and (2). I believe that the difference between (2) and (3) is very much greater."
[82] While mathematically true, we are not particularly sympathetic to the idea of infinite species survival. First, because it does not seem very plausible for there to be an *infinite* number of future humans. Second, we know that something like the heat death of the universe is an inevitable fate, regardless of how advanced future civilizations may be.

however, would be a greatly important insight for studies in philosophy and futurology.

Before concluding, we wish to raise another issue that we are yet to encounter in any of the prominent DA literature. Let us assume that Leslie's version of the DA is indeed correct. Is it *really* that surprising to find out that human civilization will probably only exist for more than a few additional millennia? Did you, prior to discovering the DA, think that we will live on for millions of more years? Granted, the main takeaway from (at least Leslie's) DA is not the specific prediction, rather, it is just the fact that we should adjust our empirical predictions (however pessimistic they may already be) upwards, and by a magnifying factor, once conditionalizing on our birth rank. That, we agree, is extraordinarily interesting and worth further academic investigation.

V. CONCLUSION

Throughout this paper we have purposefully refrained from taking a strong position on whether we think the DA is valid or not. The reasons for this are threefold. First, the vast graveyard of debunked DA objections gives us very good reason to be sceptical if we feel like we have discovered the holy grail of DA objections. Bostrom (2013: 109) notes that he has "encountered over a hundred objections against the DA… many of them mutually inconsistent. It is as if the DA is so counterintuitive that people reckon every criticism must be valid." Leslie (1993: 19) reiterates this caution: "given 20 seconds, many people believe they have found crushing objections… at least a dozen times, I too dreamed up what seemed a crushing refutation. Be suspicious of such refutations, no matter how proud you may be of them!"[83] This dissertation has been written very much with these words resonating in the back of our mind.

Second, some of the conceptual issues raised in this paper are still being actively fleshed out today. For instance, two working papers at the Global Priorities Institute in Oxford are looking at several issues we encountered. Teruj Thomas (GPI working paper, 2022) makes an

---

[83] An interesting recent paper by Turchin (2019) attempts a *Meta-Doomsday-Argument by* statistically analysing hundreds of DA objections and then using the concept of logical uncertainty to give predictive power as a probability that the argument is valid. Turchin (2019) arrives at a figure of 0.42 chance that the DA is correct, concluding that "given the large uncertainty in adding researchers to the list, this could be approximated as around even credence in DA or against it."

argument that we cannot use Lewis's Principal Principle to update our credence based on knowledge of chances unless we have first settled indexical facts about who, when, and where we are. He argues that we cannot start out with a chance-based prediction for doom which we then update based on self-locating information. Rather, it is only *after* we have the relevant self-locating information that we can use the chances to update. This is related to the point previously made by Poundstone (2019a; 183) regarding adjusting our forecast of doom upon learning one's temporal location. Meanwhile, Andreas Mogensen (GPI working paper, 2020) presents an argument as to why we in fact may be justified in thinking we are indeed exceptionally early in the grand totality of humanity. Mogensen effectively agrees with Leslie's formulation but argues that we have good reasons for thinking that we are actually amongst the first 5 percent of humans to exist.

Third, and finally, we feel that the contextual nature of the topic at hand means we should be *extra* cautious when presenting novel interpretations of the DA. It is difficult to imagine something more perilously delicate in the study of philosophy than the question of our ultimate (collective) demise. We feel that because the subject matter is incredibly sensitive, it is permissible to proceed with prudence. Of course, we acknowledge that this rationally defensible position comes at the cost of taking a strong stance on the validity of the DA. Nonetheless, we remain optimistic that our remarks on the possibility of simulating entire world timelines may shed some much-needed light on the DA puzzle.

*****

This dissertation has provided a critical assessment of the use of thought-experiments in DA literature with a focus on three distinct examples: Leslie's Urns, Eckhardt's Shooting Room, and Northcott's Asteroid. In Part II we argued that our best chance at debunking the DA is not by way of constructing thought-experiments, rather, we need to move beyond analogies and attempt to test the DA empirically. Part III found Sober's critique of the underlying sampling assumptions valid for Gott's DA, however inconclusive when considering Leslie's DA. We then suggested that if we could actually simulate real-life worlds, we would be able to better test the DA, as well as obtain better answers to some difficult questions that were raised by the aforementioned thought-experiments. Lastly, we looked at the broader implications of the DA. Even if the DA is correct, we concluded that it need not be as apocalyptic as previously imagined.

REFERENCES

Aaronson, S. (2013). *Quantum Computing Since Democritus*. Cambridge University Press.

Adams, T. (2007). Sorting out the anti-doomsday arguments: A reply to Sowers. *Mind, 116*(462), 269-273.

Bartha, P., & Hitchcock, C. (1999a). No one knows the date or the hour: An unorthodox application of rev. Bayes's theorem. *Philosophy of Science, 66*, S339-S353.

Bartha, P., & Hitchcock, C. (1999b). The shooting-room paradox and conditionalizing on measurably challenged sets. *Synthese, 118*(3), 403-437.

Bertrand, J. (1889). *Calcul des Probabilités*. New York: Chelsea, 3rd Ed., 1960.

Bostrom, N. (1997). Investigations into the Doomsday argument. *preprint*, 359-387.

Bostrom, N. (1999). The doomsday argument is alive and kicking. *Mind, 108*(431), 539-551.

Bostrom, N. (2000). Observer-relative chances in anthropic reasoning?. *Erkenntnis, 52*(1), 93-108.

Bostrom, N. (2001a). The Doomsday Argument Adam & Eve, UN++, and Quantum Joe. *Synthese, 127*(3), 359-387.

Bostrom, N. (2001b). "A Super-Newcomb Problem." *Analysis*.

Bostrom, N. (2002a). *Anthropic bias: Observation selection effects in science and philosophy*. Routledge.

Bostrom, N. (2002b). Beyond the Doomsday Argument: Reply to Sowers and Further Remarks.

Bostrom, N. (2003). Are we living in a computer simulation?. *The philosophical quarterly, 53*(211), 243-255.

Bostrom, N. (2004) A Doomsday argument Primer. https://anthropic-principle.com/primer1

Bostrom, N., & Ćirković, M. (2003). The doomsday argument and the self-indication assumption: reply to Olum. *The Philosophical Quarterly, 53*(210), 83-91.

Bostrom, N. (2008). The doomsday argument. *Think, 6*(17-18), 23-28.

Bostrom, N. (2007). Sleeping Beauty and Self-Location: A Hybrid Model. *Synthese, 157*(1), 59–78.

Bradley, D. (2005). No doomsday argument without knowledge of birth rank: A Defense of Bostrom. *Synthese, 144*(1), 91-100.

Bradley, D. (2007). *BAYESIANISM AND SELF-LOCATING BELIEFS or TOM BAYES MEETS JOHN PERRY* (Doctoral dissertation, STANFORD UNIVERSITY).

Bradley, D. (2012). Four problems about self-locating belief. *The Philosophical Review, 121*(2)

Bradley, D. & Fitelson, B. (2003). Monty Hall, Doomsday and confirmation. *Analysis, 63*(1), 23-31.

Briggs, W. (2016). *Uncertainty: the soul of modelling, probability & statistics*. Springer.

Carter, B. (1983). The anthropic principle and its implications for biological evolution. *Philosophical Transactions of the Royal Society of London. Series A, Mathematical and Physical Sciences, 310*(1512), 347-363.

Caves, C. M. (2000). Predicting future duration from present age: A critical assessment. *Contemporary Physics, 41*(3), 143-153.

Ciampanelli, P. (2015). *12 Oldest Animal Species on Earth.* 26 Feb. 2015, https://mom.com/momlife/17976-12-oldest-animal-species-earth.

Costas, E., de Alcañíz, J. G., & Lopez-Rodas, V., (2020). Groundbreaking predictions about COVID-19 pandemic duration, number of infected and dead: a novel mathematical approach never used in epidemiology. *Medrxiv.* doi:10.1101/2020.08.05.20168781

Cushman, M. (2019). Anthropic Indexical Sampling and Implications for The Doomsday Argument.

Dawkins, R., & Krebs, J. R. (1979). Arms races between and within species. *Proceedings of the Royal Society of London. Series B. Biological Sciences, 205*(1161), 489-511.

Delahaye, J.-P. (1996). "Recherche de modèles pour l'argument de l'Apocalypse de Carter-Leslie". Unpublished manuscript.

Dicke, R. (1961). "Dirac's Cosmology and Mach's Principle". *Nature*. 192 (4801): 440–41

Dieks, D. (1992). Doomsday--or: The dangers of statistics. *The Philosophical Quarterly (1950), 42*(166), 78-84.

Dieks, D. (2007). Reasoning about the future: Doom and Beauty. *Synthese, 156*(3), 427-439.

Eckhardt, W. (1993). Probability theory and the Doomsday argument. *Mind, 102*(407), 483-488.

Eckhardt, W. (1997). A shooting-room view of doomsday. *The Journal of philosophy, 94*(5), 244

Eckhardt, W. (2013). DOOMSDAY!. In *Paradoxes in Probability Theory*. Springer, Dordrecht.

Elga, A. (2000). Self-locating belief and the Sleeping Beauty problem. *Analysis, 60*(2), 143-147.

Eliazar, I. (2017). Lindy's law. *Physica A: Statistical Mechanics and its Applications, 486*, 797-805.

Everett, H. (1957). The Many-Worlds Interpretation of Quantum Mechanics,. 221 Thesis. *Princeton University, 222*.

Franceschi, P. (1998). A solution to the Doomsday Argument. *Canadian Journal of Philosophy, 28*(2).

Franceschi, P. (1999). The Doomsday Argument and Hempel's Problem. *Canadian Journal of Philosophy, 29*(1).

Franceschi, P. (2007). Compléments pour une théorie des distorsions cognitives. *Journal de thérapie comportementale et cognitive, 17*(2), 84-88.

Franceschi, P. (2009). A Third Route to the Doomsday Argument. *Journal of Philosophical Research, 34*, 263-278.

Franceschi, P. (2012). On the Disanalogy in the Simulation Argument. Unpublished manuscript https://citeseerx.ist.psu.edu/viewdoc/download?doi=10.1.1.693.5221&rep=rep1&type=pdf

Friederich, S. (2016). Self-location and causal context. *grazer philosophische studien*, *93*(2), 232-258.

Gelman, A., & Robert, C. P. (2013). "Not only defended but also applied": The perceived absurdity of Bayesian inference. *The American Statistician*, *67*(1), 1-5.

Goodman, S. N. (1983). *Fact, fiction, and forecast*. Harvard University Press. p. 74

Goodman, S. N. (1994). Future-Prospects Discussed. *Nature*, *368*(6467), 106-107.

Gott, J. R. (1993). Implications of the Copernican principle for our future prospects. *Nature*, *363*(6427), 315-319.

Gott, J. R. (1994). Future prospects discussed. *Nature*, *368*.

Gott, J. R. (1997). "A Grim Reckoning." *New Scientist,* November 15.

Gerig, A. (2012). The Doomsday Argument in Many Worlds. *arXiv preprint arXiv:1209.6251*.

Greenberg, M. (1999). Apocalypse Not Just Now. *London Review of Books*, *21*(13).

Hájek, A. (2007). The reference class problem is your problem too. *Synthese*, *156*(3), 563-585.

Hacking, I. (1964). On the foundations of statistics. *The British Journal for the Philosophy of Science*, *15*(57), 1-26.

Hanson, R. (1998). Critiquing the Doomsday Argument. *preprint di*, *5*. http://mason.gmu.edu/~rhanson/Nodoom.html

Hempel, C. G. (1945). Studies in the Logic of Confirmation (I.). *Mind*, *54*(213), 1-26.

Hofkirchner, W., & Kreowski, H. J. (Eds.). (2021). *Transhumanism: The Proper Guide to a Posthuman Condition Or a Dangerous Idea?*. Springer.

Kopf, T., Krtous, P., & Page, D. N. (1994). Too soon for doom gloom?. *arXiv preprint gr-qc/9407002*.

Korb, K. B., & Oliver, J. J. (1999). A refutation of the doomsday argument. *Mind, 107*(426), 403-410.

Kühne, U. (2005). *Die Methode des Gedankenexperiments*, Frankfurt: Suhrkamp.

Lampton, M. (2020). Doomsday: A Response to Simpson's Second Question. *arXiv preprint arXiv:2003.00132*.

Lerner, E. (1993) 'Horoscopes for Humanity?', *The New York Times*, July 14.

Leslie, J. (1989a). Is the end of the world nigh?. *The Philosophical Quarterly (1950-), 40*(158), 65-72.

Leslie, J. (1989b) 'No inverse gambler's fallacy in cosmology', *Mind,* April, 269–72.

Leslie, J. (1992a). The doomsday argument. *The Mathematical Intelligencer, 14*(2), 48-51.

Leslie, J. (1992b). Bayes, urns, and doomsday: A reply to Barker and Schrage. *Interchange, 23*(3), 289

Leslie, J. (1992c). Time and the anthropic principle. *Mind, 101*(403), 521-540.

Leslie, J. (1993a). Doomsday revisited. *The Philosophical Quarterly (1950-), 42*(166), 85-89.

Leslie, J. (1993b). Doom and probabilities. *Mind, 102*(407), 489-491.

Leslie, J. (1994). Testing the doomsday argument. *Journal of applied philosophy, 11*(1), 31-44.

Leslie, J. (1996). *The End of the World: the science and ethics of human extinction*. Psychology Press.

Leslie, J. (2008). Infinitely Long Afterlives and the Doomsday Argument. *Philosophy, 83*(326), 519–524.

Leslie, J. (2010). The risk that humans will soon be extinct. *Philosophy, 85*(4), 447-463.

Lewis, P. J. (2010). A note on the Doomsday Argument. *Analysis, 70*(1), 27-30.

Lewis, P. J. (2013). The doomsday argument and the simulation argument. *Synthese, 190*(18), 4009-4022.

Marinoff, L. (1994). A resolution of Bertrand's paradox. *Philosophy of Science, 61*(1), 1-24.

McCutcheon, R. G. (2018). What, Precisely, is Carter's Doomsday Argument?. Draft: http://philsci-archive.pitt.edu/15936/1/gigo9f.pdf

Mogensen, A. (2019). Doomsday rings twice. Global Priorities Institute Working Paper. https://globalprioritiesinstitute.org/wp-content/uploads/Andreas-Mogensen_Doomsday-rings-twice.pdf

Monton, B., & Kierland, B. (2006). How to predict future duration from present age. *The Philosophical Quarterly, 56*(222), 16-38.

Monton, B., & Roush, S. (2001). Gott's doomsday argument. Draft: http://philsci-archive.pitt.edu/1205/1/gott1f.pdf

Nelson, K. (2009). "How and how not to make predictions with temporal Copernicanism." *Synthese* 166.1: 91-111.

Nielsen, H. B. (1980) 'Did God Have to Fine Tune the Laws of Nature to Create Light', in I. Andric, I. Dadic and N. Zovko (eds), *Particle Physics 1980,* (Amsterdam: North Holland), pp. 125-142.

Nielsen, H.B. (1989) 'Random dynamics and relations between the number of fermion generations and the fine structure constants', *Act Physica Polonica B,* May, 427–68.

Northcott, R. (2016). A dilemma for the Doomsday Argument. *Ratio, 29*(3), 268-282.

Norton, J. D. (2010). Cosmic confusions: Not supporting versus supporting not. *Philosophy of Science, 77*(4), 501-523.

Nozick, R. (1969). Newcomb's problem and two principles of choice. In *Essays in honor of Carl G. Hempel* (pp. 114-146). Springer, Dordrecht.

O'Connell, M. (2018). *To Be a Machine: Adventures Among Cyborgs, Utopians, Hackers, and the Futurists Solving the Modest Problem of Death.* Doubleday.

Olum, K. D. (2002). The doomsday argument and the number of possible observers. *The Philosophical Quarterly*, *52*(207), 164-184.

Page, D. N. (2009). Possible anthropic support for a decaying universe: a cosmic doomsday argument. *arXiv preprint arXiv:0907.4153*.

Parfit, D. (1984). Reasons and Persons Clarendon Press Oxford.

Phillips, I. (2005). Causality and the Doomsday Argument. Unpublished manuscript.
http://www.futuristsguild.org/Causality.pdf

Pisaturo, R. (2009). Past longevity as evidence for the future. *Philosophy of Science*, *76*(1), 73-100.

Pisaturo, R. (2011). The Longevity Argument.
https://philpapers.org/rec/PISTLA

Poundstone, W. (2019a). *The Doomsday Calculation: How an Equation that Predicts the Future Is Transforming Everything We Know About Life and the Universe*. Hachette UK.

Poundstone, W. (2019b, June 27). 'doomsday' math says humanity may have just 760 years left. The Wall Street Journal. Retrieved February 2, 2022, from https://www.wsj.com/articles/doomsday-math-says-humanity-may-have-just-760-years-left-11561655839

Poundstone, W. (2019c, June 28). A math equation that predicts the end of humanity. Vox. Retrieved February 2, 2022, from https://www.vox.com/the-highlight/2019/6/28/18760585/doomsday-argument-calculation-prediction-j-richard-gott

Reese, M. (2003). *Our Final Hour: A Scientist's Warning: How Terror, Error, and Environmental Disaster Threaten Humankind's Future In This Century - On Earth and Beyond*. Basic Books.

Richmond, A. (2004). Immortality and Doomsday. *American Philosophical Quarterly*, *41*(3), 235–247.

Richmond, A.M. (2006). The doomsday argument. *Philosophical Books*, *47*(2), 129-142.

Richmond, A. M. (2008). Doomsday, Bishop Usher and simulated worlds. *Ratio, 21*(2), 201-217.

Richmond, A. M. (2017). Why doomsday arguments are better than simulation arguments. *Ratio, 30*(3), 221-238.

Sandberg, A. and Bostrom, N. (2008). Whole brain emulation: A roadmap, technical report 2008-3. Tech. Rep., Future of Humanity Institute, Oxford University

Salotti, J. M. (2022). Humanity extinction by asteroid impact. *Futures, 138*, 102933.

Selvin, A. (1975). "A problem in probability (letter to the editor)". *The American Statistician*. 29 (1)

Simpson, F. (2016). Apocalypse now? Reviving the Doomsday argument. *arXiv preprint arXiv:1611.03072.*

Shackel, N. (2008). *Paradoxes of probability.* 1 Publishing 49-66.

Shulman, C., & Bostrom, N. (2012). How hard is artificial intelligence? Evolutionary arguments and selection effects. *Journal of Consciousness Studies, 19*(7-8), 103-130.

Smith, Q. (1998). Critical Notice. *Canadian Journal of Philosophy, 28*(3), 413-434.

Smith, Q (1999). *The End of The World.* 413-434.

Sober, E. (2003). An Empirical Critique of Two Versions of the Doomsday Argument–Gott's Line and Leslie's Wedge. *Synthese, 135*(3), 415-430.

Sowers Jr, G. F. (2002). The demise of the doomsday argument. *Mind, 111*(441), 37-46.

Taleb, N. N. (2012). *Antifragile: Things that gain from disorder* (Vol. 3). Random House Incorporated.

Tännsjö, T. (1997). Doom soon?. *Inquiry, 40*(2), 243-252.

Tegmark, M. (1998). The interpretation of quantum mechanics: Many worlds or many words?. *Fortschritte der Physik: Progress of Physics, 46*(6-8), 855-862.

Tegmark, M., & Bostrom, N. (2005). Is a doomsday catastrophe likely?. *Nature, 438*(7069), 754-754.

Thomas, T. (2021). Doomsday and objective chance. Global Priorities Institute Working Paper. https://globalprioritiesinstitute.org/wp-content/uploads/Thomas-Doomsday-and-Objective-Chance-Version-2.pdf

Tipler, F. J. (1994). *The physics of immortality: Modern cosmology, God, and the resurrection of the dead.* Anchor.

Turchin, A. (2010). *Structure of the global catastrophe.* URSS.

Turchin, A. (2015) *Risks of Human Extinction in the 21st Century.* RTD.

Turchin, A. (2018). Forever and Again: Necessary Conditions for "Quantum Immortality" and its Practical Implications. *Journal of Ethics and Emerging Technologies, 28*(1), 31-56.

Turchin, A. (2019). A Meta-Doomsday Argument: Uncertainty About the Validity of the Probabilistic Prediction of the End of the World.

Turchin, A. (2020). Presumptuous Philosopher Proves Panspermia.

Vos Savant, M. (1997) *The Power of Logical Thinking: Easy Lessons in the Art of Reasoning, and Hard Facts about Its Absence in Our Lives.* St. Martin's Press.

Van Valen, L. (1973). A new evolutionary law. *Evol theory, 1,* 1-30.

Vineberg, S. (2011). Paradoxes of Probability. In *Philosophy of Statistics* (pp. 713-736). N. Holland.

Weatherson, B. (2003). Doomsday and the Extinction of Baseball. Unpublished Paper. http://brian.weatherson.org/doomsday.pdf

Weintraub, R. (2009). The Doomsday Argument Revisited (a Stop in the Shooting-Room Included). *Polish Journal of Philosophy, 3*(2), 109-122.

Widstam, J. (2020). Everettian Illusion of Probability, Doomsday, and Sleeping Beauty. https://odr.chalmers.se/bitstream/20.500.12380/300787/1/Joppe%20Widstam.pdf

Wilson, A. (2020). The quantum doomsday argument. *The British Journal for the Philosophy of Science.*

Zouev, A. (2021). *The Simulation Unplugged: A Critical Assessment of Bostrom's Simulation Argument.* ZE Publishing.

Zuboff, A. (1990). One self: The logic of experience. *Inquiry*, 33(1), 39-68.

Tipler, F. J. (1994). *The physics of immortality: Modern cosmology, God, and the resurrection of the dead.* Anchor.

Turchin, A. (2010). *Structure of the global catastrophe.* URSS.

Turchin, A. (2015) *Risks of Human Extinction in the 21st Century.* RTD.

Turchin, A. (2018). Forever and Again: Necessary Conditions for "Quantum Immortality" and its Practical Implications. *Journal of Ethics and Emerging Technologies*, 28(1), 31-56.

Turchin, A. (2019). A Meta-Doomsday Argument: Uncertainty About the Validity of the Probabilistic Prediction of the End of the World.

Turchin, A. (2020). Presumptuous Philosopher Proves Panspermia.

Vos Savant, M. (1997) *The Power of Logical Thinking: Easy Lessons in the Art of Reasoning, and Hard Facts about Its Absence in Our Lives.* St. Martin's Press.

Van Valen, L. (1973). A new evolutionary law. *Evol theory*, *1*, 1-30.

Vineberg, S. (2011). Paradoxes of Probability. In *Philosophy of Statistics* (pp. 713-736). N. Holland.

Weatherson, B. (2003). Doomsday and the Extinction of Baseball. Unpublished Paper. http://brian.weatherson.org/doomsday.pdf

Weintraub, R. (2009). The Doomsday Argument Revisited (a Stop in the Shooting-Room Included). *Polish Journal of Philosophy*, 3(2), 109-122.

Widstam, J. (2020). Everettian Illusion of Probability, Doomsday, and Sleeping Beauty. https://odr.chalmers.se/bitstream/20.500.12380/300787/1/Joppe%20Widstam.pdf

Wilson, A. (2020). The quantum doomsday argument. *The British Journal for the Philosophy of Science.*

Zouev, A. (2021). *The Simulation Unplugged: A Critical Assessment of Bostrom's Simulation Argument.* ZE Publishing.

Zuboff, A. (1990). One self: The logic of experience. *Inquiry*, 33(1), 39-68.

Tipler, F. J. (1994). *The physics of immortality: Modern cosmology, God, and the resurrection of the dead.* Anchor.

Turchin, A. (2010). *Structure of the global catastrophe.* URSS.

Turchin, A. (2015) *Risks of Human Extinction in the 21st Century.* RTD.

Turchin, A. (2018). Forever and Again: Necessary Conditions for "Quantum Immortality" and its Practical Implications. *Journal of Ethics and Emerging Technologies*, 28(1), 31-56.

Turchin, A. (2019). A Meta-Doomsday Argument: Uncertainty About the Validity of the Probabilistic Prediction of the End of the World.

Turchin, A. (2020). Presumptuous Philosopher Proves Panspermia.

Vos Savant, M. (1997) *The Power of Logical Thinking: Easy Lessons in the Art of Reasoning, and Hard Facts about Its Absence in Our Lives.* St. Martin's Press.

Van Valen, L. (1973). A new evolutionary law. *Evol theory*, 1, 1-30.

Vineberg, S. (2011). Paradoxes of Probability. In *Philosophy of Statistics* (pp. 713-736). N. Holland.

Weatherson, B. (2003). Doomsday and the Extinction of Baseball. Unpublished Paper. http://brian.weatherson.org/doomsday.pdf

Weintraub, R. (2009). The Doomsday Argument Revisited (a Stop in the Shooting-Room Included). *Polish Journal of Philosophy*, 3(2), 109-122.

Widstam, J. (2020). Everettian Illusion of Probability, Doomsday, and Sleeping Beauty.

https://odr.chalmers.se/bitstream/20.500.12380/300787/1/Joppe%20Widstam.pdf

Wilson, A. (2020). The quantum doomsday argument. *The British Journal for the Philosophy of Science.*

Zouev, A. (2021). *The Simulation Unplugged: A Critical Assessment of Bostrom's Simulation Argument.* ZE Publishing.

Zuboff, A. (1990). One self: The logic of experience. *Inquiry*, 33(1), 39-68.

Tipler, F. J. (1994). *The physics of immortality: Modern cosmology, God, and the resurrection of the dead.* Anchor.

Turchin, A. (2010). *Structure of the global catastrophe.* URSS.

Turchin, A. (2015) *Risks of Human Extinction in the 21st Century.* RTD.

Turchin, A. (2018). Forever and Again: Necessary Conditions for "Quantum Immortality" and its Practical Implications. *Journal of Ethics and Emerging Technologies*, 28(1), 31-56.

Turchin, A. (2019). A Meta-Doomsday Argument: Uncertainty About the Validity of the Probabilistic Prediction of the End of the World.

Turchin, A. (2020). Presumptuous Philosopher Proves Panspermia.

Vos Savant, M. (1997) *The Power of Logical Thinking: Easy Lessons in the Art of Reasoning, and Hard Facts about Its Absence in Our Lives.* St. Martin's Press.

Van Valen, L. (1973). A new evolutionary law. *Evol theory*, 1, 1-30.

Vineberg, S. (2011). Paradoxes of Probability. In *Philosophy of Statistics* (pp. 713-736). N. Holland.

Weatherson, B. (2003). Doomsday and the Extinction of Baseball. Unpublished Paper. http://brian.weatherson.org/doomsday.pdf

Weintraub, R. (2009). The Doomsday Argument Revisited (a Stop in the Shooting-Room Included). *Polish Journal of Philosophy*, 3(2), 109-122.

Widstam, J. (2020). Everettian Illusion of Probability, Doomsday, and Sleeping Beauty. https://odr.chalmers.se/bitstream/20.500.12380/300787/1/Joppe%20Widstam.pdf

Wilson, A. (2020). The quantum doomsday argument. *The British Journal for the Philosophy of Science*.

Zouev, A. (2021). *The Simulation Unplugged: A Critical Assessment of Bostrom's Simulation Argument*. ZE Publishing.

Zuboff, A. (1990). One self: The logic of experience. *Inquiry*, 33(1), 39-68.

Tipler, F. J. (1994). *The physics of immortality: Modern cosmology, God, and the resurrection of the dead*. Anchor.

Turchin, A. (2010). *Structure of the global catastrophe*. URSS.

Turchin, A. (2015) *Risks of Human Extinction in the 21st Century*. RTD.

Turchin, A. (2018). Forever and Again: Necessary Conditions for "Quantum Immortality" and its Practical Implications. *Journal of Ethics and Emerging Technologies*, 28(1), 31-56.

Turchin, A. (2019). A Meta-Doomsday Argument: Uncertainty About the Validity of the Probabilistic Prediction of the End of the World.

Turchin, A. (2020). Presumptuous Philosopher Proves Panspermia.

Vos Savant, M. (1997) *The Power of Logical Thinking: Easy Lessons in the Art of Reasoning, and Hard Facts about Its Absence in Our Lives*. St. Martin's Press.

Van Valen, L. (1973). A new evolutionary law. *Evol theory*, 1, 1-30.

Vineberg, S. (2011). Paradoxes of Probability. In *Philosophy of Statistics* (pp. 713-736). N. Holland.

Weatherson, B. (2003). Doomsday and the Extinction of Baseball. Unpublished Paper. http://brian.weatherson.org/doomsday.pdf

Weintraub, R. (2009). The Doomsday Argument Revisited (a Stop in the Shooting-Room Included). *Polish Journal of Philosophy*, *3*(2), 109-122.

Widstam, J. (2020). Everettian Illusion of Probability, Doomsday, and Sleeping Beauty. https://odr.chalmers.se/bitstream/20.500.12380/300787/1/Joppe%20Widstam.pdf

Wilson, A. (2020). The quantum doomsday argument. *The British Journal for the Philosophy of Science*.

Zouev, A. (2021). *The Simulation Unplugged: A Critical Assessment of Bostrom's Simulation Argument*. ZE Publishing.

Zuboff, A. (1990). One self: The logic of experience. *Inquiry*, 33(1), 39-68.

APPENDIX A – AN ANTHOLOGY OF DA THOUGHT-EXPERIMENTS (1950 – 2022)

#	Thought-Experiment Name	Original Paper	Core Concepts Involved	Additional References
1	Sleeping Beauty Paradox*	Zuboff (1990)	Self-locating belief; Self-indication Assumption; Anthropic Principle	Elga (2000); Lewis (2001); Bostrom (2002a: 194), (2002); Dieks (2007)
2	Geometric Incubator	Cushman (2019)	Self-Sampling Assumption; Self-Indication Assumption;	
3	The Presumptuous Philosopher	Bostrom (2002a)	Self-Indication Assumption (argues against); Self-Sampling Assumption	Bostrom (2002a: 125); Monton (2003); Turchin (2020); Mogensen (2019)
4	The Four-Buildings Experiment	Bostrom (1997)	No-outsider Requirement	
5	God's Coin Toss	Leslie (1996)	Asymmetrical Information; Self-Indication Assumption;	Olum (2002), Bostrom (2000), Gerig (2012), Bostrom (2008), Bartha & Hitchcock (1999b), Franceschi (2002), Phillips (2005)
6	The Devil's Existence	Gerig (2012)	Many Worlds Hypothesis; Multiverse	
7	The Incubator	Bostrom (2001)	Self-Indication Assumption	Bostrom (2002); Phillips (2005)
8	Adam and Eve (3 variations)	Bostrom (2001)	Self-Sampling Assumption; Newcombe's Problem	Bostrom & Cirkovic (2003); Turchin (2018)
10	UN*	Bostrom (2001)	Self-Sampling Assumption; Casual Decision Theory;	Turchin (2018)
11	Asteroid (3 variations)	Northcott (2016)	Trumping	
12	Two Urns	Leslie (1990)	Explaining the DA; Reference Class; No-Outsider requirement	Leslie (1992a, 1992b, 1993a, 1993b, 1996, 1997); Korb & Oliver (1998); Franceschi (2009); Dieks (1992); Somers (2002); Bostrom (1999; 2001; 2002a); Adams (2007)
13	The Shooting Room Paradox	Leslie (1992)	Determinism; Old-evidence Problem; Multi-class Randomness; Fallacious Reasoning;	(Leslie (1994; 1996; 1997; 2008); Eckhardt (1997; 2013); Wientraub (2009); Bartha & Hitchcock (1999a); Bostrom (2013);
14	The Small Room and the Large Room	Leslie (1994)	DA basics; Many Worlds Quantum Theory; Determinism	Leslie (1996: 244); Smith (1998: 413-434)
15	Quantum Joe	Bostrom (2001)	Self-Sampling Assumption; the Principal Principle;	Bostrom (2013); Phillips (2005); McCutcheon (2019)
16	Merchant's Thumb Principle	Leslie (1989)	Dangers of over-relying on Odds; Probability theory	Leslie (1996, 207); Leslie (1997)
17	London or Little Puddle	Leslie (1993:191)	Probability theory; Self-Sampling Assumption	Franceschi (2007); Franceschi (2009)
18	Monty Hall Problem*	Selvin (1975) formal exposition: vos Savant (1992)	Problems of Bayesian Inference; Probability Paradox; Observation Selection Effects;	Bradley & Fitelson (2003); Lewis (2010); Wilson (2020); Gelman & Robert (2013); Richmond (2008)
19	Amnesia Chamber	Bostrom (1996)	DA without knowledge of birth rank;	Somers (2002); Bostrom (2002b)

*Indicates that the thought-experiment was designed independent of the DA and was later adapted to be relevant by others (see additional references).

#	Name	Source	Keywords	References
20	Emeralds	Leslie (1996: 222)	Self-Sampling Assumption; Random Sampling Requirement	Leslie (1997); Somers (2002); Bostrom (2002: 62)
21	Newcomb's Paradox*	Nozick (1969) (informally created by William Newcomb)	SSA; Decision Theory	Bostrom (2001a; 2001b); Turchin (2018: 18)
22	Raven Paradox (Hempel's Problem)*	Hempel (1945)	Intuition; Inductive Logic	Franceschi (1999; 2009)
23	Two Geysers	Monton & Kierland (2006)	Lindy's Law; Copernican Principle;	Nelson (2009)
24	GRUE / new riddle of induction	Goodman (1983)	Moving Goalpost DA; Problems of induction	Bradley (2007); Franceschi (2016); Morgensen (2019)
25	Standard Raffle	Leslie (1992)	Bayesian Reasoning; Basics of DA	Leslie (1996: 197); Briggs (2016: Ch.1); Poundstone (2019a; Ch.2)
26	Girl in Windowless Room	Leslie (1996: 219)	Old Evidence Problem;	Franceschi (2007)
27	Two Genetic Batches	Leslie (1996: 222)	DA basics; Self-sampling; Observer-relative chances	Bostrom (1999)
28	Dr Green or Dr Black	Leslie (1996: 227)	DA basics; Bayes Theorem	Bartha & Hitchcock (1999b)
29	Dungeon	Bostrom (2002a)	Self-Sampling Assumption	Bostrom (2013, 2001); Friederich (2016); Franceschi (2012)
30	Lazy Adam	Bostrom (2001)	Self-Sampling Assumption	Bostrom (2013: 143-144)
31	$1 Billion Atom	Leslie (1996)	Repeatability; Bayesian inference	Poundstone (2019a)
32	Quantum Suicide Machine	Tegmark (1998)	Self-Sampling	Wilson (2020); Turchin (2017)
33	The End of Galaxy	Olum (2002a)	DA basics	Turchin (2018); Phillips (2005)
34	The Fine-Tuning Argument*	Dicke (1961)	Indexical beliefs; Self-Sampling Assumption	Bostrom (2001, 2007); Bradley (2005); Wilson (2020); Page (2009)
35	Everett's Many Worlds*	Everett (1957)	Quantum DA;	Widstam (2020); Wilson (2020)
36	Betting Room	Eckhardt (1993; 1997)	Determinism; Old-evidence Problem; Multi-class Randomness;	Weatherson (2003); Aaronson (2013)
37	Three Thousand Weeks	Bostrom (2007)	Self-locating belief; Self-indication Assumption; Anthropic Principle	-
38	Cro-Magnon Man	Bostrom (2013)	DA reasoning	-
39	Baby Paradox	Delahaye (1996)	Reference Class Type II; Gott's DA	Korb and Oliver (1999: 405)

EXAMINER FEEDBACK

75%

This dissertation has many virtues, and I enjoyed reading it. It partially defends the Doomsday Argument (DA) from various thought experiments, arguing that a final verdict on DA can only be reached empirically. It concludes that the jury is still out, and it offers its own suggestions about how to make further progress. The dissertation shows thorough knowledge of a large and intricate literature, using this to construct in an Appendix a highly useful and – so far as I know – entirely novel table of different thought experiments from the literature and their various implications. Throughout, the dissertation displays engaging energy and enthusiasm. It is intricately but clearly structured. It argues convincingly against Goff's version of DA, but also that Leslie's version is more difficult to take down. It offers a positive proposal of its own at the end, namely that we might – at least in principle – empirically investigate DA via simulations of the future. But while it is good to offer a positive proposal, I wasn't yet convinced that such simulations are really feasible and really probative, or that that they really sidestep the challenges to DA highlighted earlier in the dissertation. Generally, those challenges – especially the reference class and trumping problems – are arguably more serious for DA than the dissertation concedes. But it argues its corner, and overall, there was good work here.

www.ingramcontent.com/pod-product-compliance
Lightning Source LLC
Chambersburg PA
CBHW041303240426
43661CB00011B/1004